Machinery's Handbook
Made Easy

Edward T. Janecek

Industrial Press
New York, New York

Library of Congress Cataloging-in-Publication Data
Janecek, Edward Machinery's Handbook Made Easy / Ed Janecek.
p. cm. Includes bibliographical references and index.
ISBN 978-0-8311-3448-8 (soft cover)
  1. Machine design–Handbooks, manuals, etc. 2. Mechanical
engineering–Handbooks, manuals, etc.

A complete catalog record of this book is available from the Library of Congress

INDUSTRIAL PRESS INC.
989 Avenue of the Americas
New York, NY 10018

Sponsoring Editor: John Carleo
Interior Text and Cover Design: Janet Romano
Developmental Editor: Robert Weinstein
Acquisitions Editor: Christopher Conty

1 2 3 4 5 6 7 8 9 10

Table of Contents

Table of Contents

Table of Contents

Section 7: Manufacturing Processes   🔍   161

Units Covered in this Section with: Navigation Assistant 🔍

Section 8: Fasteners   165

Units Covered in this Section

Units Covered in this Section with: Navigation Assistant 🔍

Table of Contents

Table of Contents

Cross-Reference Table

for

Machinery's Handbook 28th, and
Machinery's Handbook 29th Editions

	Machinery's Handbook Made Easy	Machinery's Handbook 29th	Machinery's Handbook 28th
Section 1: Mathematics	Pages 1–18	Pages 3–146	Pages 3–153
Shop Recommended: Numbers, Fractions, and Decimals	Pages 3–5	Page 3, Table 1	Page 3, Table 1
Shop Recommended: Geometrical Propositions	Pages 6–9	Pages 56–65	Pages 55–64
Shop Recommended: Solution of Right-Angled Triangles	Pages 10–12	Page 98	Page 97
Section 2: Mechanics and Strength of Materials	Pages 19–22	Pages 147–368	Pages 155–369
Section 3: Properties, Treatment, and Testing of Materials	Pages 23–46	Pages 369–607	Pages 370–606
Shop Recommended: System of Designating Carbon and Alloy Steels	Page 27	Page 401, Table 3	Page 401, Table 3
Shop Recommended: Application of Steels	Page 28	Pages 409–410	Pages 409–410
Shop Recommended: Classification, Approximate Compositions, and Properties Affecting Selection of Tool and Die Steels	Page 29	Pages 441–442, Table 4	Pages 441–442, Table 4
Shop Recommended: Standard Stainless Steels—Typical Compositions	Page 30	Pages 406–407, Table 6	Pages 406–407 Table 6
Shop Recommended: Properties and Applications of Cast Coppers and Copper Alloys	Page 33	Pages 515–518, Table 2	Pages 514–517, Table 2

continued on next page

Cross-Reference Table

	Machinery's Handbook Made Easy	Machinery's Handbook 29th	Machinery's Handbook 28th
Shop Recommended: Classification of Copper and Copper Alloys	Page 34	Page 514	Page 513
Shop Recommended: Properties of Plastics	Page 36	Page 551	Page 550
Shop Recommended: Bad and Good Designs in Assembling Plastics with Metal Fasteners	Page 38	Page 599, Figure 26	Page 598, Figure 26
Shop Recommended: Machining Plastics	Pages 39–40	Page 598	Pages 597–599
Shop Recommended: Drill Designs Used for Drilling Plastics	Page 41	Page 600	Page 599
Shop Recommended: Speeds and Feeds for Drilling Holes of .25 to .375 Diameter in Various Thermoplastics	Page 42	Page 601, Table 8	Page 600, Table 8
Shop Recommended: Tapping and Threading of Plastics	Page 43	Page 601	Page 600
Shop Recommended: Speeds and Number of Teeth for Sawing Plastics Materials with High—Carbon Steel Saw Blades	Page 44	Page 602, Table 9	Page 601, Table 9
Shop Recommended: Milling of Plastics	Page 44	Page 602	Page 601
Shop Recommended: Other Machining Techniques	Page 45	Page 602	Page 601
Section 4: Dimensioning, Gaging, and Measuring	Pages 47–85	Pages 608–753	Pages 609–726
Shop Recommended: Drafting Practices	Pages 49–51	Page 610, Table 2	Page 609, Table 2
Shop Recommended: Geometric Dimensioning and Tolerancing	Page 53	Pages 613 text and 618, Figure 10b	Pages 612 text and 617, Table 10b
Shop Recommended: Geometric Control Symbols	Page 55	Page 614, Table 5	Page 613, Table 5
Shop Recommended: Basic Dimensions and Definitions	Pages 58–60	Page 615–619 text	Page 615–618 text
Shop Recommended: Datum Targets	Pages 61–62	Pages 619–620	Pages 618–619

Cross-Reference Table

	Machinery's Handbook Made Easy	Machinery's Handbook 29th	Machinery's Handbook 28th
Shop Recommended: Positional Tolerance	Pages 64–65	Page 620–621	Page 619–620
Shop Recommended: Allowance for Force Fits	Pages 66–68	Page 630	Page 629
Shop Recommended: Surface Texture Symbols and Construction	Page 77	Page 742	Page 718
Shop Recommended: Application of Surface Texture Symbols	Page 77	Page 743, Figure 8	Page 719, Figure 8
Shop Recommended: Surface Roughness Produced by Common Production Methods	Page 79	Page 739, Table 1	Page 715, Table 1
Shop Recommended: Lay Symbols	Page 80	Page 745, Table 5	Page 721, Table 5
Shop Recommended: Application of Surface Texture Values to Symbol	Page 81	Page 746, Table 6	Page 722, Table 6
Section 5: Tooling and Toolmaking	Pages 87–120	Pages 754–1003	Pages 730–974
Shop Recommended: Indexable Insert Holder Application Guide	Page 95	Pages 770–772, Table 3b	Pages 746–748, Table 3b
Shop Recommended: Oversize Diameters in Drilling	Page 96	Page 897	Page 873
Shop Recommended: Straight Shank Chucking Reamers-Straight Flutes, Wire Gage Sizes	Page 97	Page 858	Page 834
Shop Recommended: Straight Shank Chucking Reamers-Straight Flutes, Letter Sizes, Decimal Sizes	Page 98	Page 859	Page 835
Shop Recommended: ANSI Straight Shank Twist Drills	Pages 102–103	Pages 868–875, Table 1	Pages 844–851, Table 1
Shop Recommended: Tap Drills and Clearance Drills for Machine Screws with American National Thread Form	Page 108	Page 2030, Table 4	Page 1934, Table 4
Shop Recommended: Characteristics of a Common Thread	Page 109	Page 1808, Figure 1	Page 1713, Figure 1

continued on next page

Cross-Reference Table

	Machinery's Handbook Made Easy	Machinery's Handbook 29th	Machinery's Handbook 28th
Shop Recommended: Tool Wear and Sharpening; Cratering, Cutting Edge Chipping, Deformation, Surface Finish, Sharpening Twist Drills, Tool Geometry, Drill Point Thinning	Pages 111–117	Pages 996–1003	Pages 967–974
Section 6: Machining Operations	Pages 121–160	Pages 1004–1323	Pages 975–1263
Shop Recommended: Selecting Cutting Conditions	Page 126	Page 1013	Page 984
Shop Recommended: Cutting Speed Formulas	Page 126	Page 1015	Page 986
Shop Recommended: Cutting Speeds and Equivalent RPM for Drills of Number and Letter Sizes	Page 129	Page 1016	Page 987
Shop Recommended: Principle Speed and Feed Tables	Page 130	Page 1021	Page 992
Shop Recommended: Combined Feed/Speed Portion of the Tables; Turning	Pages 132–134	Pages 1026–1029, Table 1	Pages 993–998, Table 1
Shop Recommended: Cutting Speeds and Feeds for Milling Plain Carbon and Alloy Steel	Pages 135–136	Page 1044, Table 11	Page 1015, Table 11
Shop Recommended: Feeds and Speeds for Drilling, Reaming, and Threading Plain Carbon and Alloy Steels	Page137	Pages 1060–1062, Table 17	Pages 1031–1033, Table 17
Shop Recommended: Cutting Speeds and Feeds for Milling Stainless Steels	Page 139	Page 1049, Table 14	Page 1020, Table 14
Shop Recommended: Common Faults and Possible Causes in Surface Grinding	Page 148	Page 1267, Table 4	Page 1198, Table 4
Shop Recommended: Rectangular Coordinate System; Absolute Coordinates	Page 150	Page 1280, Figures 1 and 2	Page 1225, Figures 1 and 2
Shop Recommended: Incremental Motions	Page 150	Page 1280, Figure 3	Page 1225, Figure 3
Shop Recommended: Typical Milling G-Codes	Page 152	Page 1286, Table 2	Page 1231, Table 2
Shop Recommended: M-Codes	Page 153	Page 1287, Table 3	Page 1232, Table 3

Cross-Reference Table

	Machinery's Handbook Made Easy	Machinery's Handbook 29th	Machinery's Handbook 28th
Shop Recommended: Degrees of Accuracy Expected with NC Machine Tools	Page 155	Page 1322, Table 1	Page 1218, Table 1
Section 7: Manufacturing Processes	Pages 161–164	Pages 1328–1516	Pages 1264–1421
Section 8: Fasteners	Pages 165–184	Pages 1517–1801	Pages 1422–1707
Shop Recommended: Drill and Counterbore Size for Socket Head Cap Screws	Page 170	Page 1683, Table 5	Page 1589, Table 5
Shop Recommended: American National Standard Hexagon and Spline Socket Flat Countersink	Page 171	Page 1684, Table 6	Page 1590, Table 6
Shop Recommended: Hexagon and Spine Socket Set Screws	Page 173	Page 1693, Table 15	Page 1599, Table 15
Shop Recommended: American National Standard Hexagon Socket Head Shoulder Screw	Page 175	Page 1686, Table 8	Page 1592, Table 8
Shop Recommended: American National Standard Hardened Ground Machine Dowel Pins	Page 177	Page 1749, Table 1	Page 1655, Table 1
Shop Recommended: Retaining Rings	Pages 180–182	Pages 1778–1784, Tables 10, 11, 12, 14	Pages 1686–1690, Tables 10, 11,12, 14
Shop Recommended: British Fasteners	Page 183	Page 1626 text	Page 1532 text
Section 9: Threads and Threading	Pages 185–204	Pages 1802–2121	Pages 1708–2026
Shop Recommended: Basic Profile of UN and UNF Screw Threads	Page 189	Page 1807, Figure 1	Page 1713, Figure 1
Shop Recommended: Standard Series and Selected Combinations-Unified Screw Threads	Page 190	Page 1817, Table 3	Page 1723, Table 3
Shop Recommended: Suggested Tap Drill Sizes for Internal Dryseal Pipe Threads	Page 192	Page 1964, Table 8	Page 1869, Table 8
Shop Recommended: Dimensions over Wires of Given Diameter	Page 194	Page 1997	Page 1902

continued on next page

Cross-Reference Table

	Machinery's Handbook Made Easy	Machinery's Handbook 29th	Machinery's Handbook 28th
Shop Recommended: Tap Drill Sizes for Threads of American National Form	Page 196	Page 2029, Table 3	Page 1934, Table 3
Shop Recommended: Simple, Compound, Differential, and Block Indexing	Page 201	Page 2079 text	Pages 1984–2026 Text
Section 10: Gears, Splines and Shafts	Pages 205–208	Pages 2125–2309	Pages 2027–2214
Section 11: Machine Elements	Pages 209–228	Pages 2314–2649	Pages 2219–2553
Shop Recommended: Key Size Versus Shaft Diameter	Page 214	Page 2472, Table 1	Page 2385, Table 1
Shop Recommended: Depth Control Values S and T for Shaft and Hub	Page 215	Page 2473, Table 2	Pages 2386 Table 2
Shop Recommended: ANSI Standard Plain and Gib Head Keys	Page 216	Page 2475, Table 3	Page 2388, Table 3
Shop Recommended: ANSI Standard Fits for Parallel and Taper Keys	Page 217	Page 2476, Table 4	Page 2389, Table 4
Shop Recommended: Set Screws for Use Over Keys	Page 218	Page 2477, Table 7	Page 2390, Table 7
Shop Recommended: ANSI Keyseat Dimensions for Woodruff Keys	Page 219	Page 2480, Table 10	Page 2393, Table 10
Shop Recommended: Finding Depth of Keyseat and Distance from Top of Key to Bottom of Shaft	Page 221	Page 2483, Table 11	Page 2397, Table 11
Section 12: Measuring Units	Pages 229–242	Pages 2652–2700	Pages 2555–2604
Shop Recommended: U.S. and Metric System Conversions	Page 232	Page 2661	Page 2565
Shop Recommended: Circular and Angular Measure Conversion Factors	Page 233	Page 2662, Table 2	Pages 2566, Table 2
Shop Recommended: Fractional Inch to Decimal Inch and Millimeter	Page 234	Page 2664, Table 7	Page 2568, Table 7
Shop Recommended: Millimeter to Inch Conversion	Page 235	Page 2665, Tables 8a and 8b	Page 2569, Tables 8a and 8b

Cross-Reference Table

	Machinery's Handbook Made Easy	Machinery's Handbook 29th	Machinery's Handbook 28th
Shop Recommended: Mixed Fractional Inches to Millimeters Conversion for 0 to 41 inches in 1/64-Inch Increments	Page 236	Page 2666, Table 10	Page 2570, Table 10
Shop Recommended: Decimal of an Inch to Millimeter Conversion	Page 237	Page 2668, Table 11	Page 2572, Table 11
Shop Recommended: Millimeters to Inches Conversion	Page 238	Page 2670, Table 12	Page 2574, Table 12
Shop Recommended: Surface texture	Page 239	Page 743	Page 719
Shop Recommended: Micrometers to Microinch Conversion	Page 240	Page 2673, Table 13b	Page 2577, Table 13b

Preface

As an instructor for 15 years, a tool and die maker for 14, and a student for over 35, I have been relying on the priceless knowledge captured within the pages of *Machinery's Handbook* for a long time. What I like most about the Handbook is its complete attention to every detail; it truly is a one-stop-shop for everything related to the processes and procedures related to manufacturing. But this strength can also be its greatest weakness. As an apprentice, I can remember searching through the pages of the "Bible" as it was known in the shop, for the correct drill to use for tapping a hole. I was overwhelmed with enough information to actually *produce* a drill. I am confident that, as an individual concerned specifically with machining processes and procedures, I am not alone in my occasional challenges in navigating *Machinery's Handbook*. It is for this reason that I have created **Machinery's Handbook Made Easy**.

I want to emphasize that this is not a book for dummies. In fact, this companion to *Machinery's Handbook* is intended for intelligent, highly skilled professionals, or professionals in training, who recognize that the real genius in their work is in the details. Furthermore, they have an appreciation for the role that best practices and procedures play in their pursuit of excellence. However, the aim of this book is quite simple: to make your life easier.

Whether you are a high school student enrolled in technical courses, a college or technical school student studying machine tool operation or machine tool theory, a mechanical engineer, or a professional working in a machine shop, I am confident that this book will save you time, money, and an occasional headache. The contents are organized in an intuitive, easy-to-follow manner that is consistent with the existing layout of *Machinery's Handbook*. In fact, I envision **Machinery's Handbook Made Easy** being open right alongside your Handbook at all times as a guide to maximum efficiency.

I don't believe in testing math skills that could be accomplished with a simple calculator or quizzing you on unnecessary vocabulary that could be looked up in the Handbook at any time. Instead, I am confident that you will find the *Apply It* feature at the end of each section to be a challenging test of your ability to apply what you have just learned to the real world. Recognizing another strength of *Machinery's Handbook*, I have carefully selected a set of *Shop Recommended* diagrams that apply to the given unit and could otherwise be easily overlooked due to the sheer volume of important content.

If you have ever puzzled over *Machinery's Handbook*, as I have, as a treasure chest of wisdom without a key, then I present you with the key in the form of this book. What you decide to do with it is up to you, but I for one have very high expectations.

Edward Janecek

How to Use This Book

Machinery's Handbook Made Easy

In this "How to Use" section of **Machinery's Handbook Made Easy**, *Machinery's Handbook 29* is expressed in *italics* and **Machinery's Handbook Made Easy** (this book) is in **bold italics**.

Machinery's Handbook 29 (MH29) is described as: "A reference book for the Mechanical Engineer, Designer, Manufacturing Engineer, Draftsman, Toolmaker, and Machinist." Currently in its 29th edition, the *Handbook* has been in continuous publication since 1914. As seen by the copyright dates below, new editions are published about every three or four years.

How *Machinery's Handbook 29* is Set Up

Machinery's Handbook 29 has:

- One Table of Contents
- Twelve Chapters
- One General Index
- Notes pages

Machinery's Handbook 29 is also available on CD. *Machinery's Handbook 29* CD has:

- One Table of Contents
- Twelve Chapters
- Three Indexes:
  - The General Index
  - The Index of Standards
  - The Index of Interactive Equations

Machinery's Handbook Made Easy is designed to save you time and money. It is organized in an intuitive, easy-to-follow manner which will help you find answers and solve common problems encountered in the machine shop. Throughout **Machinery's Handbook Made Easy,** references to the page numbers, tables, and text are matched with *Machinery's Handbook 29.*

How to Use This Book

The "Shop Recommended" feature of **Machinery's Handbook Made Easy** reproduces actual tables and other useful sources of information taken directly from *Machinery's Handbook 29*. There is a table in the front of **Machinery's Handbook Made Easy** that converts the page numbers for "Shop Recommended" to the 28th edition of *Machinery's Handbook*.

Machinery's Handbook 29 is divided into twelve major parts. The twelve parts are described in the Table of Contents and are represented on thumb tabs on the edge of the pages for quick access. The *Handbook* does not label these parts with chapter numbers or sections numbers. **Machinery's Handbook 29, Made Easy** assigns Section numbers to the twelve major parts of the *Handbook* for easy identification:

Machinery's Handbook Made Easy
The Twelve Major Sections

Section 1	**Mathematics**
Section 2	**Mechanics and Strength of Materials**
Section 3	**Properties, Treatment, and Testing of Materials**
Section 4	**Dimensioning, Gaging, and Measuring**
Section 5	**Tooling and Toolmaking**
Section 6	**Machining Operations**
Section 7	**Manufacturing Processes**
Section 8	**Fasteners**
Section 9	**Threads and Threading**
Section 10	**Gears, Spines, and Cams**
Section 11	**Machine Elements**
Section 12	**Measuring Units**

Within each major Section there are a number of subtopics. *Machinery's Handbook 29* does not label the subtopics within the Sections. **Machinery's Handbook Made Easy** assigns *Units* to the subtopics for easy identification. The Units coincide with the subtopics in each Section of *Machinery's Handbook 29* and are represented in **Machinery's Handbook 29, Made Easy** with letters of the alphabet.

As shown in
Machinery's Handbook 29

As shown in
Machinery's Handbook Made Easy

Table of Contents with Sections and Units

Machinery's Handbook 29	Machinery's Handbook 29, Made Easy
Mathematics	*Section 1 Mathematics*
Numbers, Fractions, and Decimals	• Unit A **Numbers, Fractions, and Decimals**
Algebra and Equations	• Unit B **Algebra and Equations**
Geometry	• Unit C **Geometry**
Solution of Triangles	• Unit D **Solution of Triangles**
Matrices	• Unit E **Matrices**

How to Use This Book

Manufacturing Data Analysis	• Unit F	**Manufacturing Data Analysis**
Engineering Economics	• Unit G	**Engineering Economics**

Mechanics and Strength of Materials	*Section 2 Mechanics and Strength of Materials*	
Mechanics	• Unit A	**Mechanics**
Velocity, Acceleration, Work, and Energy	• Unit B	**Velocity, Acceleration, Work, and Energy**
Strength of Materials	• Unit C	**Strength of Materials**
Properties of Bodies	• Unit D	**Properties of Bodies**
Beams	• Unit E	**Beams**
Columns	• Unit F	**Columns**
Plates, Shells, and Cylinders	• Unit G	**Plates, Shells, and Cylinders**
Shafts	• Unit H	**Shafts**
Springs	• Unit I	**Springs**
Disc Springs	• Unit J	**Disc Springs**
Fluid Mechanics	• Unit K	**Fluid Mechanics**

Properties, Treatment, and Testing of Materials	*Section 3 Properties, Treatment, and Testing of Materials*	
The Elements, Heat, Mass, and Weight	• Unit A	**The Elements, Heat, Mass, and Weight**
Properties of Wood, Ceramics,Plastics,Metals	• Unit B	**Properties of Wood, Ceramics, Plastics, Metals**
Standard Steels	• Unit C	**Standard Steels**
Tool Steels	• Unit D	**Tool Steels**
Hardening, Tempering, and Annealing	• Unit E	**Hardening, Tempering, and Annealing**
Non-ferrous Alloys	• Unit F	**Non-ferrous Alloys**
Plastics	• Unit G	**Plastics**

Dimensioning, Gaging, and Measuring	*Section 4 Dimensioning, Gaging, and Measuring*	
Drafting Practices	• Unit A	**Drafting Practices**
Allowances and Tolerances for Fits	• Unit B	**Allowances and Tolerances for Fits**
Measuring Instruments and Inspection Methods	• Unit C	**Measuring Instruments and Inspection Methods**
Surface Texture	• Unit D	**Surface Texture**

Tooling and Toolmaking	*Section 5 Tooling and Toolmaking*	
Cutting Tools	• Unit A	**Cutting Tools**
Cemented Carbides	• Unit B	**Cemented Carbides**
Forming Tools	• Unit C	**Forming Tools**
Milling Cutters	• Unit D	**Milling Cutters**
Reamers	• Unit E	**Reamers**
Twist Drills and Counterbores	• Unit F	**Twist Drills and Counterbores**
Taps	• Unit G	**Taps**
Standard Tapers	• Unit H	**Standard Tapers**
Arbors, Chucks, and Spindles	• Unit I	**Arbors, Chucks, and Spindles**
Broaches and Broaching	• Unit J	**Broaches and Broaching**
Files and Burrs	• Unit K	**Files and Burrs**
Tool Wear and Sharpening	• Unit L	**Tool Wear and Sharpening**

continued on next page

How to Use This Book

Machinery's Handbook 29	Machinery's Handbook 29, Made Easy
Machining Operations	**Section 6 Machining Operations**
Cutting Speeds and Feeds	• Unit A **Cutting Speeds and Feeds**
Speed and Feed Tables	• Unit B **Speed and Feed Tables**
Estimating Speeds and Machining Power	• Unit C **Estimating Speeds and Machining Power**
Micromachining	• Unit D **Micromachining**
Machine Econometrics	• Unit E **Machine Econometrics**
Screw Machine Feeds and Speeds	• Unit F **Screw Machine Feeds and Speed**
Machining Non-Ferrous Metals and Non-Metallic Materials	• Unit G **Machining Non-Ferrous Metals and Non-Metallic Materials**
Grinding Feeds and Speeds	• Unit H **Grinding Feeds and Speeds**
Grinding and Other Abrasive Processes	• Unit I **Grinding and Other Abrasive Processes**
CNC Numerical Control Programming	• Unit J **CNC Numerical Control Programming**
Manufacturing Processes	**Section 7 Manufacturing Processes**
Punches, Dies, and Presswork	• Unit A **Punches, Dies, and Presswork**
Electrical Discharge Machining	• Unit B **Electrical Discharge Machining**
Iron and Steel Castings	• Unit C **Iron and Steel Castings**
Soldering and Brazing	• Unit D **Soldering and Brazing**
Welding	• Unit E **Welding**
Lasers	• Unit F **Lasers**
Finishing Operations	• Unit G **Finishing Operations**
Fasteners	**Section 8 Fasteners**
Torque and Tension in Fasteners	• Unit A **Torque and Tension in Fasteners**
Inch Threaded Fasteners	• Unit B **Inch Threaded Fasteners**
Metric Threaded Fasteners	• Unit C **Metric Threaded Fasteners**
Helical Coil Screw Threaded Inserts	• Unit D **Helical Coil Screw Threaded Inserts**
British Fasteners	• Unit E **British Fasteners**
Machine Screws and Nuts	• Unit F **Machine Screws and Nuts**
Cap Screws and Set Screws	• Unit G **Cap Screws and Set Screws**
Self-Threading Screws	• Unit H **Self-Threading Screws**
T-Slots, Bolts, and Nuts	• Unit I **T-Slots, Bolts, and Nuts**
Rivets and Riveted Joints	• Unit J **Rivets and Riveted Joints**
Pins and Studs	• Unit K **Pins and Studs**
Retaining Rings	• Unit L **Retaining Rings**
Wing Nuts, Wing Screws, and Thumb Screws	• Unit M **Wing Nuts, Wing Screws, and Thumb Screws**
Nails, Spikes, and Wood Screws	• Unit N **Nails, Spikes, and Wood Screws**
Threads and Threading	**Section 9 Threads and Threading**
Screw Thread Systems	• Unit A **Screw Thread Systems**
Unified Screw Threads	• Unit B **Unified Screw Threads**
Calculating Thread Dimensions	• Unit C **Calculating Thread Dimensions**
Metric Screw Threads	• Unit D **Metric Screw Threads**
Acme Screw Threads	• Unit E **Acme Screw Threads**
Buttress Threads	• Unit F **Buttress Threads**
Whitworth Threads	• Unit G **Whitworth Threads**

Pipe and Hose Threads	• Unit H **Pipe and Hose Threads**
Other Threads	• Unit I **Other Threads**
Measuring Screw Threads	• Unit J **Measuring Screw Threads**
Tapping and Thread Cutting	• Unit K **Tapping and Thread Cutting**
Thread Rolling	• Unit L **Thread Rolling**
Thread Grinding	• Unit M **Thread Grinding**
Thread Milling	• Unit N **Thread Milling**
Simple, Compound, Differential, and Block Indexing	• Unit O **Simple, Compound, Differential, and Block Indexing**
Gears, Splines , and Cams	*Section 10 Gears, Splines, and Cams*
Gears and Gearing	• Unit A **Gears and Gearing**
Hypoid and Bevel Gearing	• Unit B **Hypoid and Bevel Gearing**
Worm Gearing	• Unit C **Worm Gearing**
Helical Gearing	• Unit D **Helical Gearing**
Other Gear Types	• Unit E **Other Gear Types**
Checking Gear Sizes	• Unit F **Checking Gear Sizes**
Gear Materials	• Unit G **Gear Materials**
Spines and Serrations	• Unit H **Spines and Serrations**
Cams and Cam Design	• Unit I **Cams and Cam Design**
Machine Elements	*Section 11 Machine Elements*
Plain Bearings	• Unit A **Plain Bearings**
Ball, Roller, and Needle Bearings	• Unit B **Ball, Roller, and Needle Bearings**
Lubrication	• Unit C **Lubrication**
Couplings, Clutches, and Brakes	• Unit D **Couplings, Clutches, and Brakes**
Keys and Keyways	• Unit E **Keys and Keyways**
Flexible Belts and Sheaves	• Unit F **Flexible Belts and Sheaves**
Transmission Chains	• Unit G **Transmission Chains**
Ball and Acme Leadscrews	• Unit H **Ball and Acme Leadscrews**
Electric Motors	• Unit I **Electric Motors**
Adhesives and Sealants	• Unit J **Adhesives and Sealants**
O-Rings	• Unit K **O-Rings**
Rolled Steel, Wire, and Sheet Metal	• Unit L **Rolled Steel, Wire, and Sheet Metal**
Shaft Alignment	• Unit M **Shaft Alignment**
Measuring Units	*Section 12 Measuring Units*
Symbols and Abbreviations	• Unit A **Symbols and Abbreviations**
Measuring Units	• Unit B **Measuring Units**
U.S. System and Metric System Conversions	• Unit C **U.S. System and Metric System Conversions**

Helpful Recurring Features

Text Boxes and Navigation Assistant

Watch for recurring features throughout *Machinery's Handbook Made Easy.* For example, helpful information may be found inside of a text box or accompanied by the

Navigation Assistant symbol: . The Navigation Assistant helps to locate information in *Machinery's Handbook MH29* and in *Machinery's Handbook Made Easy*. Here is an example:

Navigation Assistant There is one General Index in the back of MH29, containing thousands of entries.

Using the General Index

In the following example, the technician needs to clarify the meaning of a welding symbol shown on an engineering drawing. The key word in this example is "welding." Under "Welding" in the General Index, the word "symbol" is found:

Index navigation path and key words:

— welding/symbol/page 1477

symbol 1477
  arrow side 1482
  bead type back 1481
  bevel groove 1480
  built up surface 1481
  electron beam 1480, 1484
  fillet 1480–1481
  intermittent fillet 1482
  letter designations 1478
  melt thru weld 1484
  plug groove 1480
  process 1478

> Under the heading "symbol" in the General Index are a number of related topics

How to Use This Book

Table 1 shows the result of the search, found on page 1477 in *Machinery's Handbook 29.*

Table 1. Basic Weld Symbols

Groove Weld Symbols							
Square	Scarf[a]	V	Bevel	U	J	Flare V	Flare bevel

Other Weld Symbols							
Fillet	Plug or slot	Spot or projection	Seam	Back or backing	Surfacing	Flange	
						Edge	Corner

Units within the Sections

Each section of this guide covers the individual units that are found in the *Handbook.* Some units are developed further with the Navigation Assistant. Others, Like Section 11: Machine Elements, Unit E: *Keys and Keyseats*, are approached in a different way, including **Key Terms, Learning Objectives,** and an Introduction, followed by **Design Considerations** and **Machining Methods.** These elements are seen in the following example, drawn from Section 11, Unit E.

Part of **Section 11: Machine Elements,** Unit E: Keys and Keyseats

How to Use This Book

Key Terms

Keyseat	Wire EDM
Keyway	Keyseat Alignment
Woodruff Key	Tolerance
Broach	Clearance Fit
Keyseating machine	Interference Fit
Broach	

Learning Objectives and Key Terms are common themes in:
Machinery's Handbook 29, Made Easy

Learning Objectives:

After studying this unit you should be able to:

- Select the proper size key for a given shaft diameter.
- Interpret the size of a Woodruff key by its part number.
- State the three methods of machining a keyway.
- Calculate allowable tolerances for machining keyways and keyseats.
- State the difference between a clearance fit and an interference fit.

Introduction

A key is a small component of an assembly that is used to align shafts, hubs, pulleys, handles, cutters, and gears. Metal keys lie in a shallow groove with part of the key in each component. They are used to transmit motion through shafts by eliminating slippage. Keys are square, rectangular (*flat*), or semi-circular (*woodruff*). Like electrical fuses, keys are designed to fail, or shear, before more expensive components are damaged.

In an assembly with a shaft and a hub, the *keyway* is the groove in the hub and the *keyseat* is the groove in the shaft (see Figure 11.1). The fits between the components of a keyed assembly are closely controlled.

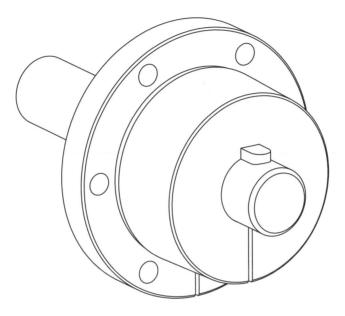

Figure 11.1 Keyed assembly of shaft and hub

Navigation Assistant

Not all *Units* within the major *Sections* are covered the same way. Some *Units* are covered with the **Navigation Assistant** only. These *Units* give navigation tips for popular topics. In the following example, Units shown from Section 11: Machine Elements are covered with the **Navigation Assistant**. Unit A: "Plain Bearings" has a short introduction followed by navigation tips only. The navigation tips make finding information in the General Index easier.

How to Use This Book

Unit A: Plain Bearings pages 2314–2363

Plain Bearings prevent wear by providing sliding contact between mating surfaces.
This unit describes the three classes of Plain Bearings and gives characteristics, applications, advantages, and disadvantages of *Plain Bearings*.

> Units covered with the **Navigation Assistant** may have a short introduction

Popular Searches in: Plain Bearings, pages 2314–2364

Index navigation path and key words:

— bearings/plain/pages 2314—2364
— bearings/plain/classes of/page 2314
— bearings/plain/greases/page 2325
— bearings/plain/materials/page 2356

> Unit A: Plain Bearings and all other Units in this Section are covered by the Navigation Assistant only.

Shop Recommended

The "Shop Recommended" feature of ***Machinery's Handbook Made Easy*** refers to actual tables and other sources of information taken directly from *Machinery's Handbook 29*. In the following example, Table 1 (Key Size Versus Shaft Diameter) has been copied from page 2472. This table and other tables throughout the book have been selected because of their relevance to basic machine shop practices. Below Table 1, text boxes explain how to interpret useful information. The ***Shop Recommended*** feature appears throughout ***Machinery's Handbook Made Easy***.

Shop Recommended page 2472, Table 1

Table 1. Key Size Versus Shaft Diameter *ANSI B17.1-1967 (R2008)*

Nominal Shaft Diameter		Nominal Key Size			Normal Keyseat Depth	
			Height, *H*		*H*/2	
Over	To (Incl.)	Width, *W*	Square	Rectangular	Square	Rectangular
$5/16$	$7/16$	$3/32$	$3/32$	…	$3/64$	…
$7/16$	$9/16$	$1/8$	$1/8$	$3/32$	$1/16$	$3/64$
$9/16$	$7/8$	$3/16$	$3/16$	$1/8$	$3/32$	$1/16$
$7/8$	$1\ 1/4$	$1/4$	$1/4$	$3/16$	$1/8$	$3/32$
$1\ 1/4$	$1\ 3/8$	$5/16$	$5/16$	$1/4$	$5/32$	$1/8$
$1\ 3/8$	$1\ 3/4$	$3/8$	$3/8$	$1/4$	$3/16$	$1/8$
$1\ 3/4$	$2\ 1/4$	$1/2$	$1/2$	$3/8$	$1/4$	$3/16$
$2\ 1/4$	$2\ 3/4$	$5/8$	$5/8$	$7/16$	$5/16$	$7/32$
$2\ 3/4$	$3\ 1/4$	$3/4$	$3/4$	$1/2$	$3/8$	$1/4$
$3\ 1/4$	$3\ 3/4$	$7/8$	$7/8$	$5/8$	$7/16$	$5/16$
$3\ 3/4$	$4\ 1/2$	1	1	$3/4$	$1/2$	$3/8$
$4\ 1/2$	$5\ 1/2$	$1\ 1/4$	$1\ 1/4$	$7/8$	$5/8$	$7/16$
$5\ 1/2$	$6\ 1/2$	$1\ 1/2$	$1\ 1/2$	1	$3/4$	$1/2$
Square Keys preferred for shaft diameters above this line; rectangular keys, below						
$6\ 1/2$	$7\ 1/2$	$1\ 3/4$	$1\ 3/4$	$1\ 1/2$[a]	$7/8$	$3/4$
$7\ 1/2$	9	2	2	$1\ 1/2$	1	$3/4$
9	11	$2\ 1/2$	$2\ 1/2$	$1\ 3/4$	$1\ 1/4$	$7/8$

> This is the nominal depth of the *keyseat*. The actual machined depth is shown in MHB 29, page 2473; Table 2

> *Nominal size* is the size used for the general identification of the shaft size.

> *Width* of key recommended for the shaft shown on the left.

> Square keys have the same *width W* and *height H*. Rectangular keys are sometimes called "flat keys" and are recommended for larger shaft diameters.

How to Use This Book

Safety First

Look for **Safety First** text boxes throughout **Machinery's Handbook Made Easy.** The machine shop environment is potentially dangerous. Accidents are caused by one, or a combination of the following:

- Personal Factors
- Unsafe Conditions
- Unsafe Practices

Personal Factors

A personal factor is something in one's personal life that is causing a distraction. An example of a personal factor is a worker worried about the health of a loved one.

Unsafe Conditions

An example of an unsafe condition is a fork lift with faulty brakes. Unsafe conditions refer to equipment, machines, and tools that are in a state of disrepair, broken, or missing important components or guards.

Unsafe Practices

Unsafe practices and personal factors are the leading cause of accidents in industry. Using a machine with the guards removed is an example of an unsafe practice.

This is an example of a **Safety First** text box:

> ### Safety First
>
> **Never operate a machine with the safety guards removed.**

Analyze, Evaluate, & Implement

To practice the concepts presented in the twelve Sections of **Machinery's Handbook Made Easy**, the **Analyze, Evaluate, & Implement** feature provides learning activities throughout the book.

- **Analyze**   the material presented.
- **Evaluate**   what you have learned by thinking critically.
- **Implement**   the lesson with an activity that reinforces the lesson.

How to Use This Book

This example of **Analyze, Evaluate, & Implement** is from Section 8; Fasteners:

> *Analyze, Evaluate, & Implement*

Make a table listing popular sizes of socket head cap screws. For every size of socket head cap screw, list the size of the clearance drill, the diameter of the screw head, and the recommended depth of the counterbore. The counterbore depth should allow the fastener to be .03 recessed below the surface of the material. Save the table for reference.

How to use the Index in *Machinery's Handbook 29*

Use the General Index in the back of the book to locate the information you need. There is one General Index containing thousands of entries. The index is set up alphabetically.

To locate a topic in the General Index,

1. Determine the proper name of the subject you need information about.

2. Find the listing in the General Index. If several pages are listed, check each page for the information you need.

3. If your subject is not listed, try to think of another word that describes the topic.

How to Use This Book

Use the flowchart in Figure 1 to find a topic *Machinery's Handbook*.

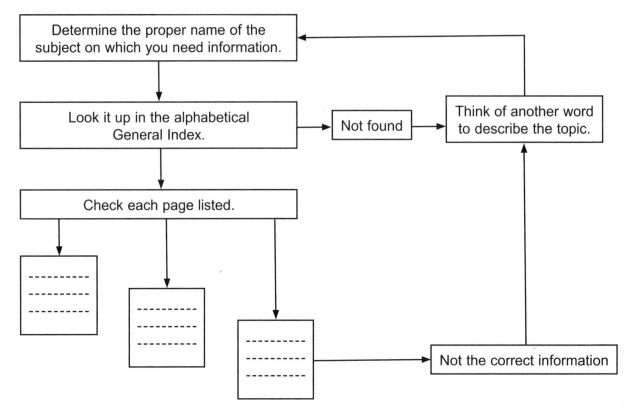

Figure 1 Finding a topic in Machinery's Handbook 29

> *Analyze, Evaluate, & Implement*

Use Machinery's Handbook 29 to locate the following:

1. Composition of SAE steels

2. Grinding wheel safety

3. Sizes of number size drills

4. Surface area of a cylinder

5. Tap drill sizes

6. Drill sizes for reaming

7. Surface texture symbols

8. Three different processes to produce a thread

9. Metric to English conversion factors

10. Speed and feed tables for turning mild steel

APPLY *IT!*

Finally, at the end of selected Sections there are a number of questions under the heading **APPLY IT!** The questions have been carefully chosen to represent the types of issues that occur in an actual manufacturing environment or machine shop. The answers to these questions are given in the back of the book.

Acknowledgements

The creation of this book has been an enlightening personal journey for me as its author. I am very proud of the result and sincerely hope that this text does what I set out for it to do, which was to make each reader's life a little easier. Of course, I never could have accomplished this alone, so I would like to take a moment to recognize those who have helped me along the way.

I would like to begin by thanking the publisher, Industrial Press. In particular, I owe a special thanks to Production Manager and Art Director Janet Romano for her helpful and friendly advice throughout the process. Thanks are also due to Robert Weinstein for his editorial guidance in crafting a more readable and better organized manuscript. I also want to recognize Christopher Conty for his detailed and expert guidance through the early stages of this project.

Next, I would like to recognize my current employer, Waukesha County Technical College, and my Associate Dean Mike Shiels. As an instructor in the Manufacturing Technology Department for the past 15 years, I have been afforded many opportunities to advance both personally and professionally through the college's training programs, workshops, and professional network. And to my students, most of whom are professionals in training who spend every working day in the field, I would like to say thank you for keeping my classroom relevant with fresh perspectives and new challenges.

Last, I would like to thank Stanek Tool where I worked for 15 years, first as an apprentice and then as a Tool & Die Maker. There, I learned the art of metal removal using proper machine shop practices. I was fortunate to learn from skilled craftsmen who stressed quality and the importance of being detail-oriented. They also taught me methods and short cuts you will not find in any book. In particular, I would like to thank Mary Wehrheim, President of Stanek Tool, for providing an environment that truly cared about its employees. It was at Stanek Tool where I first came to appreciate the process of working as part of a team, creating tools that make the products people use and depend on every day.

Section 1

MATHEMATICS

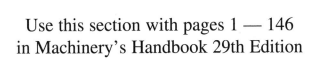

Use this section with pages 1 — 146
in Machinery's Handbook 29th Edition

SECTION 1

Navigation Overview

Units Covered in this Section

Units Covered in this Section with:
Navigation Assistant

The Navigation Assistant helps find information in the MH29 Primary Index. The Primary Index is located in the back of the book on pages 2701-2788 and is set up alphabetically by subject. Watch for the magnifying glass throughout the section for navigation hints.

May I help you?

Key Terms

Numbers	Hypotenuse
Fractions	Bisect
Improper Fraction	Perpendicular
Decimal	Sine
Reciprocal	Cosecant
Numerator	Cosine
Mixed Number	Secant
Denominator	Tangent
Proposition	Cotangent
Equilateral	Reciprocal
Isosceles	

Learning Objectives

After studying this unit you should be able to:

- Convert a fraction to decimal inch.
- Define point, line, and plain.
- State four geometric propositions.
- State the names of the sides of a right triangle.
- Compute the sine, cosine, and tangent of a given angle
- Compute the cosecant, secant, and cotangent of a given angle.
- State the Pythagorean Theorem.

Introduction

Units A, C, and D explore basic mathematical concepts that are used to solve common machine shop problems such as working with fractions and decimals, converting millimeters to inches, using right triangle trigonometry to calculate machine positions, and applying geometric propositions to aid in interpreting engineering drawings. In the machine shop, virtually all mathematical computations are done with a calculator. It is assumed that the reader is familiar with the operation of the scientific calculator.

Unit A: Numbers, Fractions, and Decimals pages 3–9

 Index navigation paths and key words:

— numbers/positive and negative/page 4

Calculations are made using *numbers, fractions,* and *decimals. Whole numbers* are the first ten numbers: 0, 1, 2, 3, 4, 5, 6, 7, 8, 9. Common fractions are made up of two parts: a *denominator,* or bottom number, and a *numerator,* or top number. Decimals are values expressed in terms of tenths, hundredths, and thousandths. In the machine shop decimals are expressed in thousandths. For example, .2 is expressed as "two hundred thousandths;" .03 is "thirty thousandths;" and .006 is "six thousandths." When decimals have four places, the fourth place is referred to as "tenths", so .0107 is pronounced "ten thousandths and seven tenths."

Rules and Definitions for Common Fractions

- A *proper fraction* is a fraction where the *numerator* (top number) is smaller than the bottom number (*denominator*).
- An *improper fraction* is a fraction where the *numerator* is greater than the denominator, for example, 7/5, 6/4, and 41/32. *Improper fractions* have values greater than 1.
- Mixed numbers are made up of a whole number and a fraction, for example, 1 1/2, 2 3/4, and 6 7/8 are mixed numbers

---

Analyze, Evaluate, & Implement

Express the following dimensions in word form:

Example: .02   <u>twenty thousandths</u>

.023 _____

.0076 _____

2.10 _____

.0401 _____

.8754 _____

Shop Recommended page 3, Table 1

NUMBERS, FRACTIONS, AND DECIMALS

Table 1. Fractional and Decimal Inch to Millimeter, Exact[a] Values

Fractional Inch	Decimal Inch	Millimeters	Fractional Inch	Decimal Inch	Millimeters
1/64	0.015625	0.396875		0.511811024	13
1/32	0.03125	0.79375	33/64	0.515625	13.096875
	0.039370079	1	17/32	0.53125	13.49375
3/64	0.046875	1.190625	35/64	0.546875	13.890625
1/16	0.0625	1.5875		0.551181102	14
5/64	0.078125	1.984375	9/16	0.5625	14.2875
	0.078740157	2	37/64	0.578125	14.684375
1/12	$0.08\overline{33}$[b]	$2.11\overline{66}$	7/12	$0.58\overline{33}$	$14.81\overline{66}$
3/32	0.09375	2.38125		0.590551181	15
7/64	0.109375	2.778125	19/32	0.59375	15.08125
	0.118110236	3	39/64	0.609375	15.478125
1/8	0.125	3.175	5/8	0.625	15.875
9/64	0.140625	3.571875		0.62992126	16
5/32	0.15625	3.96875	41/64	0.640625	16.271875
	0.157480315	4	21/32	0.65625	16.66875
1/6	$0.1\overline{66}$	$4.2\overline{33}$	2/3	$0.\overline{66}$	$16.9\overline{33}$
11/64	0.171875	4.365625		0.669291339	17
3/16	0.1875	4.7625	43/64	0.671875	17.065625
	0.196850394	5	11/16	0.6875	17.4625
13/64	0.203125	5.159375	45/64	0.703125	17.859375
7/32	0.21875	5.55625		0.708661417	18
15/64	0.234375	5.953125	23/32	0.71875	18.25625
	0.236220472	6	47/64	0.734375	18.653125
1/4	0.25	6.35		0.748031496	19
17/64	0.265625	6.746875	3/4	0.75	19.05
	0.275590551	7	49/64	0.765625	19.446875
9/32	0.28125	7.14375	25/32	0.78125	19.84375
19/64	0.296875	7.540625		0.787401575	20
5/16	0.3125	7.9375	51/64	0.796875	20.240625
	0.31496063	8	13/16	0.8125	20.6375
21/64	0.328125	8.334375		0.826771654	21
1/3	$0.\overline{33}$	$8.4\overline{66}$	53/64	0.828125	21.034375
11/32	0.34375	8.73125	27/32	0.84375	21.43125
	0.354330709	9	55/64	0.859375	21.828125
23/64	0.359375	9.128125		0.866141732	22
3/8	0.375	9.525	7/8	0.875	22.225
25/64	0.390625	9.921875	57/64	0.890625	22.621875
	0.393700787	10		0.905511811	23
13/32	0.40625	10.31875	29/32	0.90625	23.01875
5/12	$0.41\overline{66}$	$10.58\overline{33}$	11/12	$0.91\overline{66}$	$23.28\overline{33}$
27/64	0.421875	10.715625	59/64	0.921875	23.415625
	0.433070866	11	15/16	0.9375	23.8125
7/16	0.4375	11.1125		0.94488189	24
29/64	0.453125	11.509375	61/64	0.953125	24.209375
15/32	0.46875	11.90625	31/32	0.96875	24.60625
	0.472440945	12		0.984251969	25
31/64	0.484375	12.303125	63/64	0.984375	25.003125
1/2	0.5	12.7			

Decimal Inch: Numerator divided by the Denominator

Proper Fraction

Metric Equivalent given in millimeters. (Decimal Inch divided by 25.4. To convert millimeters to inches, multiply by 25.4)

[a] Table data are based on 1 inch = 25.4 mm, exactly. Inch to millimeter conversion values are exact. Whole number millimeter to inch conversions are rounded to 9 decimal places.
[b] Numbers with an overbar, repeat indefinitely after the last figure, for example $0.08\overline{33} = 0.08333...$

SECTION 1

Unit C: Geometry, pages 37–94

🔍 Index navigation paths and key words:

A *Geometric Proposition* is a figure made from points and lines that lie on a flat plain. Geometric Propositions are known to be true. An engineering drawing is an example of points and lines drawn on a flat plain that conveys precise information to the machinist.
A point is an exact location in space. A line is a series of points in space that are straight and continue infinitely in both directions. A plane is a flat surface with no thickness.

Shop Recommended pages 56–65, Geometrical Propositions

56   GEOMETRICAL PROPOSITIONS

Geometrical Propositions

Engineering drawings often use angles to convey precise information. When two angles of a triangle are shown, the remaining angle can be determined.

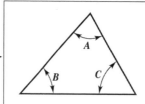

The sum of the three angles in a triangle always equals 180 degrees. Hence, if two angles are known, the third angle can always be found.

$$A + B + C = 180° \qquad A = 180° - (B + C)$$
$$B = 180° - (A + C) \qquad C = 180° - (A + B)$$

If one side and two angles in one triangle are equal to one side and similarly located angles in another triangle, then the remaining two sides and angle also are equal.

If $a = a_1$, $A = A_1$, and $B = B_1$, then the two other sides and the remaining angle also are equal.

If two sides and the angle between them in one triangle are equal to two sides and a similarly located angle in another triangle, then the remaining side and angles also are equal.

If $a = a_1$, $b = b_1$, and $A = A_1$, then the remaining side and angles also are equal.

If the three sides in one triangle are equal to the three sides of another triangle, then the angles in the two triangles also are equal.

If $a = a_1$, $b = b_1$, and $c = c_1$, then the angles between the respective sides also are equal.

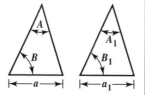
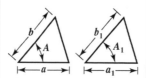
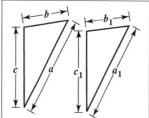

continued on next page

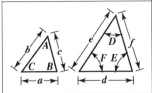	If the three sides of one triangle are proportional to corresponding sides in another triangle, then the triangles are called *similar*, and the angles in the one are equal to the angles in the other. If $a:b:c = d:e:f$, then $A = D$, $B = E$, and $C = F$.
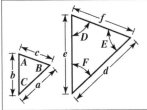	If the angles in one triangle are equal to the angles in another triangle, then the triangles are similar and their corresponding sides are proportional. If $A = D$, $B = E$, and $C = F$, then $a:b:c = d:e:f$.
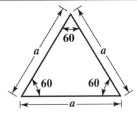	If the three sides in a triangle are equal—that is, if the triangle is *equilateral*—then the three angles also are equal. Each of the three equal angles in an equilateral triangle is 60 degrees. If the three angles in a triangle are equal, then the three sides also are equal.

Machining thirty degree and sixty degree angles is a common machine shop operation because these angles are used for leads, clearances, and wrench flats.

SECTION 1

Geometrical Propositions

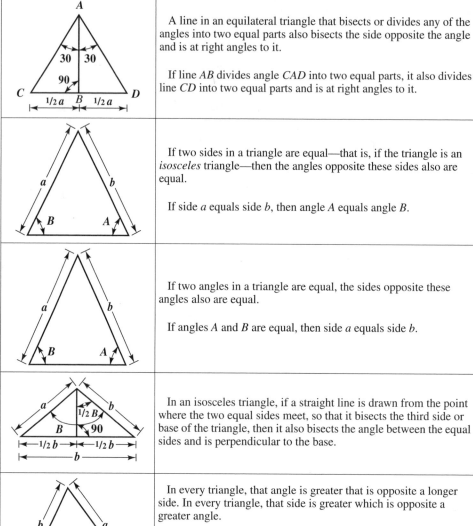

A line in an equilateral triangle that bisects or divides any of the angles into two equal parts also bisects the side opposite the angle and is at right angles to it.

If line *AB* divides angle *CAD* into two equal parts, it also divides line *CD* into two equal parts and is at right angles to it.

If two sides in a triangle are equal—that is, if the triangle is an *isosceles* triangle—then the angles opposite these sides also are equal.

If side *a* equals side *b*, then angle *A* equals angle *B*.

If two angles in a triangle are equal, the sides opposite these angles also are equal.

If angles *A* and *B* are equal, then side *a* equals side *b*.

In an isosceles triangle, if a straight line is drawn from the point where the two equal sides meet, so that it bisects the third side or base of the triangle, then it also bisects the angle between the equal sides and is perpendicular to the base.

In every triangle, that angle is greater that is opposite a longer side. In every triangle, that side is greater which is opposite a greater angle.

If *a* is longer than *b*, then angle *A* is greater than *B*. If angle *A* is greater than *B*, then side *a* is longer than *b*.

In every triangle, the sum of the lengths of two sides is always greater than the length of the third.

Side *a* + side *b* is always greater than side *c*.

In a right-angle triangle, the square of the hypotenuse or the side opposite the right angle is equal to the sum of the squares on the two sides that form the right angle.

$$a^2 = b^2 + c^2$$

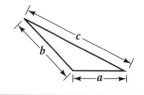

Use this formula for determining the side of a right triangle when two sides are known:
$$a^2 = b^2 + c^2$$

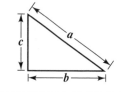

Mathematics

Geometrical Propositions

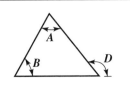	If one side of a triangle is produced, then the exterior angle is equal to the sum of the two interior opposite angles. Angle D = angle A + angle B
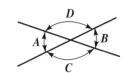	If two lines intersect, then the opposite angles formed by the intersecting lines are equal. Angle A = angle B Angle C = angle D
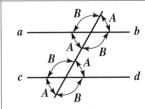	If a line intersects two parallel lines, then the corresponding angles formed by the intersecting line and the parallel lines are equal. Lines ab and cd are parallel. Then all the angles designated A are equal, and all those designated B are equal.
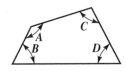	In any figure having four sides, the sum of the interior angles equals 360 degrees. $A + B + C + D$ = 360 degrees
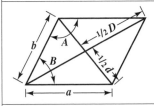	The sides that are opposite each other in a parallelogram are equal; the angles that are opposite each other are equal; the diagonal divides it into two equal parts. If two diagonals are drawn, they bisect each other.
	The areas of two parallelograms that have equal base and equal height are equal. If $a = a_1$ and $h = h_1$, then Area A = area A_1
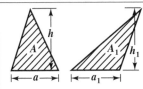	The areas of triangles having equal base and equal height are equal. If $a = a_1$ and $h = h_1$, then Area A = area A_1
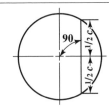	If a diameter of a circle is at right angles to a chord, then it bisects or divides the chord into two equal parts.

These propositions are useful when working with engineering drawings. See Figure 1.1.

Determine angle "X" in the drawing below

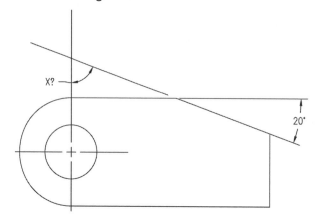

Figure 1.1

Unit D: Solution of Triangles pages 95–121

 Index navigation path and key words:

In the machine shop, right triangle trigonometry is used to solve a variety of problems such as determining the coordinates of a bolt circle. A right triangle is a triangle with a 90-degree angle. The side opposite the 90-degree angle is the *hypotenuse,* which is always the longest side of the triangle. *Sine, cosine, and tangent* are ratios that are used to solve triangle problems. *Cosecant, secant, and cotangent* are *reciprocals* of *sine, cosine, and tangent,* or values that equal 1 divided by the *sine, cosine or tangent* respectively.

Naming the Sides of a Right Triangle

The sides of a right triangle are known as the side opposite, the side adjacent, and the hypotenuse. The hypotenuse is the side opposite the 90 degree angle and is always the longest side of the triangle (see Figure 1.2). The side opposite and the side adjacent are assigned with respect to the angle referenced. For example, in triangle BAC, Side BC is the side opposite angle A. Side CA is the side opposite angle B.

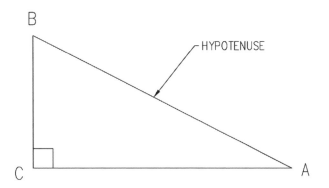

Figure 1.2 The sides and angles of a right triangle.

Sine, Cosine, and Tangent

Sine, cosine, and tangent are ratios that are used to solve triangle problems. The scientific calculator is used to determine the sine, cosine, or tangent of any angle. Calculator procedures vary based on the different makes and models, but one of the two following procedures will work with almost all. To determine the tangent of 12 degrees:

12 TAN = .212556562 or TAN 12 = .212556562

Cosecant, Secant, and Cotangent

Cosecant, secant, and cotangent are reciprocals of sine, cosine, and tangent. In other words, cotangent is one over the tangent of the angle or one divided by the tangent of the angle. To find the cotangent of 12 degrees:

12 TAN 1/X = 4.70463011

The Pythagorean Theorem

The Pythagorean Theorem is used to solve an unknown side of a right triangle when two sides are known. The hypotenuse, which is always the longest side of a right triangle, is c. The formula is:

$$a^2 + b^2 = c^2$$

Simply square the sides bc and ac and find the square root of the result.

When the hypotenuse is known, the formula changes slightly:

$$c^2 - b^2 = a^2$$

Shop Recommended page 98, Solution of Right-Angled Triangles

Solution of Right-Angled Triangles

In this right triangle, the angles are represented by the capitol letters A, B, and C. The sides are represented by the lower case letters a, b, and c.

As shown in the illustration, the sides of the right-angled triangle are designated a and b and the hypotenuse, c. The angles opposite each of these sides are designated A and B, respectively.

Angle C, opposite the hypotenuse c is the right angle, and is therefore always one of the known quantities.

Sides and Angles Known	Formulas for Sides and Angles to be Found		
Side a; side b	$c = \sqrt{a^2 + b^2}$	$\tan A = \dfrac{a}{b}$	$B = 90° - A$
Side a; hypotenuse c	$b = \sqrt{c^2 - a^2}$	$\sin A = \dfrac{a}{c}$	$B = 90° - A$
Side b; hypotenuse c	$a = \sqrt{c^2 - b^2}$	$\sin B = \dfrac{b}{c}$	$A = 90° - B$
Hypotenuse c; angle B	$b = c \times \sin B$	$a = c \times \cos B$	$A = 90° - B$
Hypotenuse c; angle A	$b = c \times \cos A$	$a = c \times \sin A$	$B = 90° - A$
Side b; angle B	$c = \dfrac{b}{\sin B}$	$a = b \times \cot B$	$A = 90° - B$
Side b; angle A	$c = \dfrac{b}{\cos A}$	$a = b \times \tan A$	$B = 90° - A$
Side a; angle B	$c = \dfrac{a}{\cos B}$	$b = a \times \tan B$	$A = 90° - B$
Side a; angle A	$c = \dfrac{a}{\sin A}$	$b = a \times \cot A$	$B = 90° - A$

In the right triangle above, side b and angle B are known. Replace the parts of the formula that are know and solved.

1

The General Index is in the back of the MH.

Unit B: Algebra and Equations pages 30–36

Algebra is the branch of mathematics that uses variables to represent numbers. The rules of algebra can be used to solve equations such as calculating RPM, determining thermal expansion of metal, and finding the weight of a material using a formula.

Index navigation paths and key words:

— algebra and equations/page 30

Unit E: Matrices pages 122–127

Logarithms and Matrices are advanced tools of mathematics used to solve complex problems. For the machinist, the use of logarithms and matrixes is unlikely. Most shop calculations are conversions from one system of measurement to another, right-triangle trigonometry, and solving formulas.

Unit F: Manufacturing Data Analysis pages 128–133

Quality is a vital part of manufacturing. The goal of today's industry is to produce products with zero defects. Controlled sampling of piece part dimensions can be represented on a chart or table to give a visual illustration of the dimensions. A pattern of results can be studied which may lead to predictions of future part quality. The practice of *gathering manufacturing data for analysis* is known as Statistical Process Control, or SPC.

Unit G: Engineering Economics pages 134–146

Engineers, managers, and purchasing agents use cost analysis techniques to reduce manufacturing cost or increase production. This area of manufacturing is known as Engineering Economics.

SECTION 1

 The Primary Index has thousands of entries.

ASSIGNMENT

List the key terms and give a definition of each.

Numbers	Denominator	Sine
Fractions	Proposition	Cosecant
Improper Fraction	Equilateral	Cosine
Decimal	Isosceles	Secant
Reciprocal	Hypotenuse	Tangent
Numerator	Bisect	Cotangent
Mixed Number	Perpendicular	Reciprocal

APPLY IT! PART 1

1. Solve for:

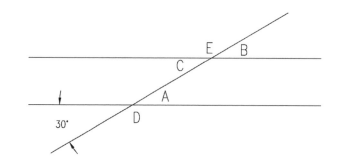

A. _____

B. _____

C. _____

D. _____

E. _____

2. Angle X = _____

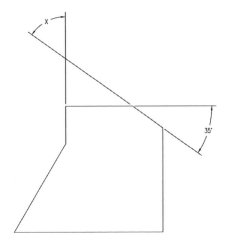

3. Angle X = _____

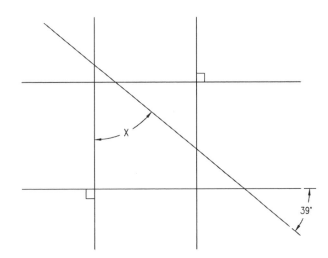

4. Angle X = _____ Angle Y = _____

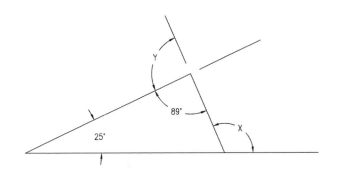

5. Angle X = _____   Angle Y = _____

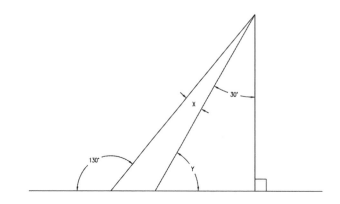

6. Angle X = _____

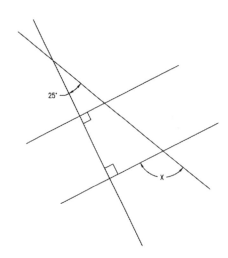

7. Angle X = _____

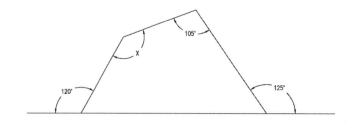

8. Angle X = _____

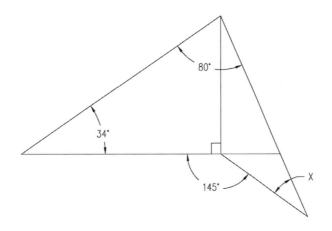

APPLY IT! PART 2

1. What is the sine of 34°?

2. Label the triangle with respect to Angle B.

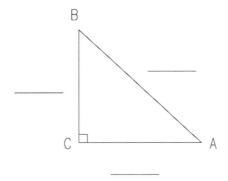

3. What is the cotangent of 30°? _____

4. What is the reciprocal of 2? _____

5. What is the cosine of 42°?_____

6. What is the secant of 72°? _____

7. A = _____   B = _____   C = _____

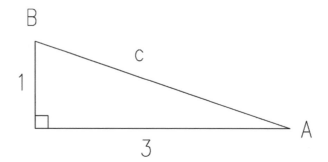

8. A = _____   B = _____

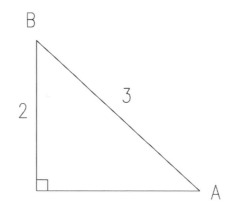

Section 2

MECHANICS and STRENGTH of MATERIALS

Use this section with pages 147 — 368
in Machinery's Handbook 29th Edition

2

Navigation Overview

Units Covered in this Section with:
Navigation Assistant

The Navigation Assistant helps find information in the MH29 Primary Index. The Primary Index is located in the back of the book on pages 2701-2788 and is set up alphabetically by subject. Watch for the magnifying glass throughout the section for navigation hints.

May I help you?

Introduction

The subject of mechanics with regard to manufacturing and engineering is very diverse. By definition, mechanics is the branch of physics concerned with the behavior of physical bodies when subjected to forces or displacements. The Section Mechanics and Strength of Materials may appear to be a collection of unrelated topics, but they do have things in common. Units A through K in this section provide information on important machine design considerations such as the properties of motion, choice of materials, strength of materials, choosing the correct spring, and formulas for work and power.

Unit A: Mechanics pages 149–174

Check these pages for information

— Terms and definitions/pages 149–153
— Force Systems/pages 153–164

Mechanics and Strength of Materials

Unit B: Velocity, Acceleration, Work, and Energy
pages 175–198

Check these pages for information

Unit C: Strength of Materials pages 199–220

Check these pages for information

Unit D: Properties of Bodies pages 221–255

Check these pages for information

Unit E: Beams pages 256–280

Check these pages for information

Unit F: Columns pages 281–287

Check these pages for information

Unit G: Plates, Shells, and Cylinders pages 288–294

2

Section 3

PROPERTIES, TREATMENT, and TESTING of MATERIALS

Use this section with pages 369 — 607
in Machinery's Handbook 29th Edition

SECTION 3

3

Navigation Overview

Units Covered in this Section

    Unit C   Standard Steels
    Unit D   Tool Steels
    Unit E   Hardening, Tempering, and Annealing
    Unit F   Nonferrous Alloys
    Unit G   Plastics

Units Covered in this Section with:
Navigation Assistant

The Navigation Assistant helps find information in the MH29 Primary Index. The Primary Index is located in the back of the book on pages 2701-2788 and is set up alphabetically by subject. Watch for the magnifying glass throughout the section for navigation hints.

May I help you?

    Unit A   Elements, Heat, Mass, and Weight
    Unit B   Properties of Wood, Ceramics, Plastics, Metals

Key Terms

Alloy	Harden Temper
Element	Anneal
Hardened	Stress Relieve
Tempered	Thermosets
Pig Iron	Thermoplastics

> ## Learning Objectives
>
> *After studying this unit you should be able to:*
>
> - State the ingredients of steel.
> - Describe six mechanical properties of tool steels.
> - List four types of stainless steels and their uses.
> - Recognize challenges in the machining of stainless steels.
> - Describe five ways to improve the machining of stainless steels.
> - Define the three stages of hardening steel.
> - State an application for annealing, normalizing, and stress relieving.

3

Introduction

Steel does not occur naturally. It is an *alloy* of iron and less than about 2% carbon. An *alloy* is two or more *elements* forming a mixture in a solid solution. The raw materials used to make steel are iron ore, coal, and limestone to make to make *pig iron*. Pig iron, iron, and steel scrap are refined by removing undesirable elements and adding desirable elements in measured amounts. Other elements can be added to the iron-carbon mixture during the manufacturing stage to create other alloys. There are several systems used to classify and identify different types of steels. The *AISI* and *SAE* numerical system uses a four— or five-digit number to identify the steel and its alloying elements. Most steel used in industry is low carbon steel having less than 0.25% carbon. Low carbon steel is also known as mild steel, machine steel, cold rolled steel, 1018, and 1020.

All types of steel fall into one of the following categories:

- Carbon Steel
- Alloy Steel
- Stainless Steel

Carbon Steels are an alloy of iron and carbon with trace amounts of impurities.

Alloy Steels contain iron, carbon and other elements to give the steel certain mechanical and chemical properties. Common alloying elements are:

- Chromium
- Molybdenum
- Nickel
- Vanadium
- Manganese
- Tungsten

SECTION 3

Stainless Steels contain chromium as a major alloying element. Stainless steels do rust and corrode like ordinary steel.

The properties of metal can be changed. By introducing heat in a controlled fashion, *hardening*, *tempering*, and *annealing* are possible. The science of heat treating is discussed in this section.

Materials other than steel are used extensively in industry. There are many types of metals that do not contain iron, known as nonferrous metal, and alloys of nonferrous metals. The plastics industry is monumental. Products made of plastic such as drink bottles, floor tiles, car fenders, and toothbrushes affect our everyday lives.

Unit C: Standard Steels pages 396–431

Carbon is the single most important alloying element in steel because very small changes in the amount of carbon result in significant changes in the properties of the steel. Low carbon steel has a carbon content of about two-hundredths of one percent. Its uses include automobile body panels, storage tanks, bridges, fence wire, and ships. Medium carbon steel has .30 to .55% carbon. Its uses include crankshafts, axles, bolts, and connecting rods. High carbon steel has .55 to about 1% carbon and is used for springs, plow blades, music wire, snap rings, and thrust washers. Low carbon steel is easy to machine and weld. High carbon steel is difficult to machine and weld.

Properties, Treatment, and Testing of Materials

Shop Recommended page 401, Table 3

Table 3. AISI-SAE System of Designating Carbon and Alloy Steels

AISI-SAE Designation[a]		Type of Steel and Nominal Alloy Content (%)
		Carbon Steels
10xx		Plain Carbon (Mn 1.00% max.)
11xx		Resulfurized
12xx		Resulfurized and Rephosphorized
15xx		Plain Carbon (Max. Mn range 1.00 to 1.65%)
		Manganese Steels
13xx		Mn 1.75
		Nickel Steels
23xx		Ni 3.50
25xx		Ni 5.00
		Nickel-Chromium Steels
31xx		Ni 1.25; Cr 0.65 and 0.80
32xx		Ni 1.75; Cr 1.07
33xx		Ni 3.50; Cr 1.50 and 1.57
34xx		Ni 3.00; Cr 0.77
		Molybdenum Steels
40xx		Mo 0.20 and 0.25
44xx		Mo 0.40 and 0.52
		Chromium-Molybdenum Steels
41xx		Cr 0.50, 0.80, and 0.95; Mo 0.12, 0.20, 0.25, and 0.30
		Nickel-Chromium-Molybdenum Steels
43xx		Ni 1.82; Cr 0.50 and 0.80; Mo 0.25
43BVxx		Ni 1.82; Cr 0.50; Mo 0.12 and 0.35; V 0.03 min.
47xx		Ni 1.05; Cr 0.45; Mo 0.20 and 0.35
81xx		Ni 0.30; Cr 0.40; Mo 0.12
86xx		Ni 0.55; Cr 0.50; Mo 0.20
87xx		Ni 0.55; Cr 0.50; Mo 0.25
88xx		Ni 0.55; Cr 0.50; Mo 0.35
93xx		Ni 3.25; Cr 1.20; Mo 0.12
94xx		Ni 0.45; Cr 0.40; Mo 0.12
97xx		Ni 0.55; Cr 0.20; Mo 0.20
98xx		Ni 1.00; Cr 0.80; Mo 0.25
		Nickel-Molybdenum Steels
46xx		Ni 0.85 and 1.82; Mo 0.20 and 0.25
48xx		Ni 3.50; Mo 0.25
		Chromium Steels
50xx		Cr 0.27, 0.40, 0.50, and 0.65
51xx		Cr 0.80, 0.87, 0.92, 0.95, 1.00, and 1.05
50xxx		Cr 0.50; C 1.00 min.
51xxx		Cr 1.02; C 1.00 min.
52xxx		Cr 1.45; C 1.00 min.
		Chromium-Vanadium Steels
61xx		Cr 0.60, 0.80, and 0.95; V 0.10 and 0.15 min
		Tungsten-Chromium Steels
72xx		W 1.75; Cr 0.75
		Silicon-Manganese Steels
92xx		Si 1.40 and 2.00; Mn 0.65, 0.82, and 0.85; Cr 0.00 and 0.65
		High-Strength Low-Alloy Steels
9xx		Various SAE grades
xxBxx		B denotes boron steels
xxLxx		L denotes leaded steels
AISI	SAE	Stainless Steels
2xx	302xx	Chromium-Manganese-Nickel Steels
3xx	303xx	Chromium-Nickel Steels
4xx	514xx	Chromium Steels
5xx	515xx	Chromium Steels

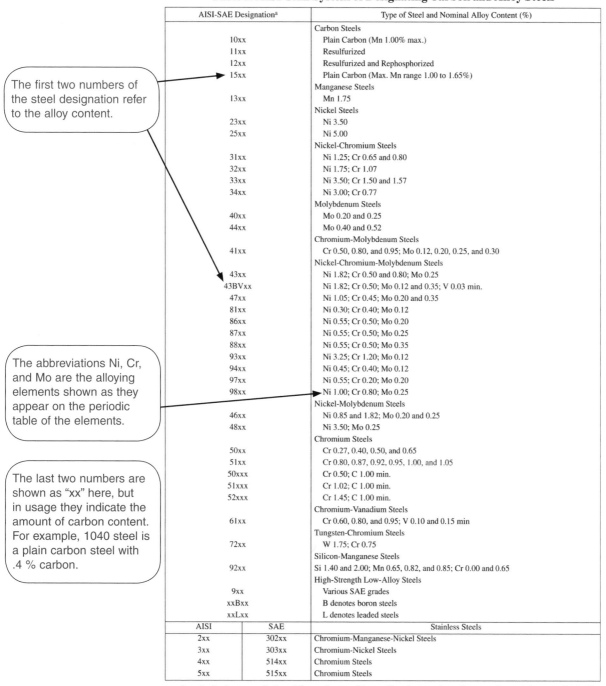

> The first two numbers of the steel designation refer to the alloy content.

> The abbreviations Ni, Cr, and Mo are the alloying elements shown as they appear on the periodic table of the elements.

> The last two numbers are shown as "xx" here, but in usage they indicate the amount of carbon content. For example, 1040 steel is a plain carbon steel with .4 % carbon.

[a] xx in the last two digits of the carbon and low-alloy designations (but not the stainless steels) indicates that the carbon content (in hundredths of a per cent) is to be inserted.

SECTION 3

Shop Recommended pages 409–410, Application of Steels

recommended for a given application, information on the characteristics of each steel listed will be found in the section *Carbon Steels* starting on page 410.

This sample of the *Application of Steels* list on page 409–410 shows an application followed by the AISI-SAE number of the recommended material.

Adapters, 1145
Agricultural steel, 1070, 1080
Aircraft forgings, 4140
Axles front or rear, 1040, 4140
Axle shafts, 1045, 2340, 2345, 3135, 3140, 3141, 4063, 4340
Ball-bearing races, 52100
Balls for ball bearings, 52100
Body stock for cars, rimmed*
Bolts and screws, 1035
Bolts
   anchor, 1040

Fan blades, 1020
Fatigue resisting 4340, 4640
Fender stock for cars, rimmed*
Forgings
   aircraft, 4140
   carbon steel, 1040, 1045
   heat-treated, 3240, 5140, 6150
   high-duty, 6150
   small or medium, 1035
   large, 1036
Free-cutting steel
   carbon, 1111, 1113

Unit D: Tool Steels pages 433–460

Tool steels have their own category because they are used to make the tools (or machines) that make the products we use every day. Tool steels are usually used in the *hardened* and *tempered* condition. *Hardening* is a hot process that changes the molecular structure of the material to make it more wear resistant and resistant to penetration. *Tempering* follows hardening to relax the structure of the steel and make it less brittle.

Tool steel has desirable *mechanical properties* such as:

- Resistance to deformation   (Hardness)
- The ability to withstand impact   (Toughness)
- Resistance to being twisted   (Torsional Strength)
- The ability to perform at high temperatures (Hot Hardness)
- Resistance to being sheared   (Shear Strength)
- The ability to resist being pulled apart   (Tensile Strength)

Properties, Treatment, and Testing of Materials

Shop Recommended pages 441–442, Table 4

The "Type of Tool Steel" column gives the name of the steel and its possible compositions. Water, Oil, and Air refer to the quenching method that is used in the heat treating process.

"Safety in Hardening" refers to metals' ability to stay straight, flat, and crack resistant during the heat treating process.

Table 4. Classification, Approximate Compositions, and Properties Affecting Selection of Tool and Die Steels
(From SAE Recommended Practice)

Type of Tool Steel	Chemical Composition[a]								Non-warping Prop.	Safety in Hardening	Tough-ness	Depth of Hardening	Wear Resistance
	C	Mn	Si	Cr	V	W	Mo	Co					
Water Hardening													
0.80 Carbon	70–0.85	b	b	b	...	...	...	...	Poor	Fair	Good[c]	Shallow	Fair
0.90 Carbon	0.85–0.95	b	b	b	...	...	...	...	Poor	Fair	Good[c]	Shallow	Fair
1.00 Carbon	0.95–1.10	b	b	b	...	...	...	...	Poor	Fair	Good[c]	Shallow	Good
1.20 Carbon	1.10–1.30	b	b	b	...	...	...	...	Poor	Fair	Good[c]	Shallow	Good
0.90 Carbon-V	0.85–0.95	b	b	b	0.15–0.35	...	...	...	Poor	Fair	Good	Shallow	Fair
1.00 Carbon-V	0.95–1.10	b	b	b	0.15–0.35	...	...	...	Poor	Fair	Good	Shallow	Good
1.00 Carbon-VV	0.90–1.10	b	b	b	0.35–0.50	...	...	...	Poor	Fair	Good	Shallow	Good
Oil Hardening													
Low Manganese	0.90	1.20	0.25	0.50	0.20[d]	0.50	...	...	Good	Good	Fair	Deep	Good
High Manganese	0.90	1.60	0.25	0.35[d]	0.20[d]	...	0.30[d]	...	Good	Good	Fair	Deep	Good
High-Carbon, High-Chromium[e]	2.15	0.35	0.35	12.00	0.80[d]	0.75[d]	0.80[d]	...	Good	Good	Poor	Through	Best
Chromium	1.00	0.35	0.25	1.40	...	...	0.40	...	Fair	Good	Fair	Deep	Good
Molybdenum Graphitic	1.45	0.75	1.00	...	...	...	0.25	...	Fair	Good	Fair	Deep	Good
Nickel-Chromium[f]	0.75	0.70	0.25	0.85	0.25[d]	...	0.50[d]	...	Fair	Good	Fair	Deep	Fair
Air Hardening													
High-Carbon, High-Chromium	1.50	0.40	0.40	12.00	0.80[d]	...	0.90	0.60[d]	Best	Best	Fair	Through	Best
5 Per Cent Chromium	1.00	0.60	0.25	5.25	0.40[d]	...	1.10	...	Best	Best	Fair	Through	Good
High-Carbon, High-Chromium-Cobalt	1.50	0.40	0.40	12.00	0.80[d]	...	0.90	3.10	Best	Best	Fair	Through	Best
Shock-Resisting													
Chromium-Tungsten	0.50	0.25	0.35	1.40	0.20	2.25	0.40[d]	...	Fair	Good	Good	Deep	Fair
Silicon-Molybdenum	0.50	0.40	1.00	...	0.25[d]	...	0.50	...	Poor[g]	Poor[h]	Best	Deep	Fair
Silicon-Manganese	0.55	0.80	2.00	0.30[d]	0.25[d]	...	0.40[d]	...	Poor[g]	Poor[h]	Best	Deep	Fair
Hot Work													
Chromium-Molybdenum-Tungsten	0.35	0.30	1.00	5.00	0.25[d]	1.25	1.50	...	Good	Good	Good	Through	Fair
Chromium-Molybdenum-V	0.35	0.30	1.00	5.00	0.40	...	1.50	...	Good	Good	Good	Through	Fair
Chromium-Molybdenum-VV	0.35	0.30	1.00	5.00	0.90	...	1.50	...	Good	Good	Good	Through	Fair
Tungsten	0.32	0.30	0.20	3.25	0.40	9.00	...	...	Good	Good	Good	Through	Fair

The reason these steels harden to different depths is because they are made up of different elements quenched with different mediums. Water quenching is the fastest, resulting in a thin layer of hardness, or "case hardened." Air quenching is slower, allowing the material to harden all the way through.

3

Stainless Steels

Stainless steel does not rust or corrode like ordinary (mild) steel. There are many different grades and types of stainless steel, but they all share chromium as an alloying element. Chromium is what makes stainless steels corrosion resistant. In addition to iron and chromium, other elements such as nickel, molybdenum, silicon, and manganese may be added to give the steel special properties. Stainless steel contains from 11% to 30% chromium. Uses of stainless steel include automotive trim, tableware, knives, aircraft parts, cooking utensils, and food processing equipment. All types of stainless steel fall into one of four broad categories:

- Martensitic
- Ferritic
- Austenitic
- Heat Resisting

These terms refer to the molecular structure of the material.

Shop Recommended pages 406–407, Table 6

406   CHEMICAL COMPOSITION OF STAINLESS STEELS

Table 6. Standard Stainless Steels — Typical Compositions

Austenitic: Contains little or no carbon. Cannot be hardened by heat treating, but will work-harden. It is non-magnetic when soft, magnetic when work-hardened. Known as "300" series.

AISI Type (UNS)	Typical Composition (%)	AISI Type (UNS)	Typical Composition (%)
Austenitic			
201 (S20100)	16-18 Cr, 3.5-5.5 Ni, 0.15 C, 5.5-7.5 Mn, 0.75 Si, 0.060 P, 0.030 S, 0.25 N	310 (S31000)	24-26 Cr, 19-22 Ni, 0.25 C, 2.0 Mn, 1.5 Si, 0.045 P, 0.030 S
202 (S20200)	17-19 Cr, 4-6 Ni, 0.15 C, 7.5-10.0 Mn, 0.75 Si, 0.060 P, 0.030 S, 0.25 N	310S (S31008)	24-26 Cr, 19-22 Ni, 0.08 C, 2.0 Mn, 1.5 Si, 0.045 P, 0.30 S
205 (S20500)	16.5-18 Cr, 1-1.75 Ni, 0.12-0.25 C, 14-15.5 Mn, 0.75 Si, 0.060 P, 0.030 S, 0.32-0.40 N	314 (S31400)	23-26 Cr, 19-22 Ni, 0.25 C, 2.0 Mn, 1.5-3.0 Si, 0.045 P, 0.030 S
301 (S30100)	16-18 Cr, 6-8 Ni, 0.15 C, 2.0 Mn, 0.75 Si, 0.045 P, 0.030 S	316 (S31600)	16-18 Cr, 10-14 Ni, 0.08 C, 2.0 Mn, 0.75 Si, 0.045 P, 0.030 S, 2.0-3.0 Mo, 0.10 N
302 (S30200)	17-19 Cr, 8-10 Ni, 0.15 C, 2.0 Mn, 0.75 Si, 0.045 P, 0.030 S, 0.10 N	316L (S31603)	16-18 Cr, 10-14 Ni, 0.03 C, 2.0 Mn, 0.75 Si, 0.045 P, 0.030 S, 2.0-3.0 Mo, 0.10 N
302B (S30215)	17-19 Cr, 8-10 Ni, 0.15 C, 2.0 Mn, 2.0-3.0 Si, 0.045 P, 0.030 S	316F (S31620)	16-18 Cr, 10-14 Ni, 0.08 C, 2.0 Mn, 1.0 Si, 0.20 P, 0.10 S min, 1.75-2.50 Mo
303 (S30300)	17-19 Cr, 8-10 Ni, 0.15 C, 2.0 Mn, 1.0 Si, 0.20 P, 0.015 S min, 0.60 Mo (optional)	316N (S31651)	16-18 Cr, 10-14 Ni, 0.08 C, 2.0 Mn, 0.75 Si, 0.045 P, 0.030 S, 2-3 Mo, 0.10-0.16 N
303Se (S30323)	17-19 Cr, 8-10 Ni, 0.15 C, 2.0 Mn, 1.0 Si, 0.20 P, 0.060 S, 0.15 Se min	317 (S31700)	18-20 Cr, 11-15 Ni, 0.08 C, 2.0 Mn, 0.75 Si, 0.045 P, 0.030 S, 3.0-4.0 Mo, 0.10 N max

Ferritic: Contains little or no carbon. Has very little alloying elements other than iron and chromium. Known as "400 series." Magnetic.

AISI Type (UNS)	Typical Composition (%)	AISI Type (UNS)	Typical Composition (%)
Ferritic			
405 (S40500)	11.5-14.5 Cr, 0.08 C, 1.0 Mn, 1.0 Si, 0.040 P, 0.030 S, 0.1-0.3 Al, 0.60 max	430FSe (S43023)	16-18 Cr, 0.12 C, 1.25 Mn, 1.0 Si, 0.060 P, 0.060 S, 0.15 Se min
409 (S40900)	10.5-11.75 Cr, 0.08 C, 1.0 Mn, 1.0 Si, 0.045 P, 0.030 S, 0.05 Ni (Ti 6 × C, but with 0.75 max)	434 (S43400)	16-18 Cr, 0.12 C, 1.0 Mn, 1.0 Si, 0.040 P, 0.030 S, 0.75-1.25 Mo
429 (S42900)	14-16 Cr, 0.12 C, 1.0 Mn, 1.0 Si, 0.040 P, 030 S, 0.75 Ni	436 (S43600)	16-18 Cr, 0.12 C, 1.0 Mn, 1.0 Si, 0.040 P, 0.030 S, 0.75-1.25 Mo (Nb + Ta 5 × C min, 0.70 max)
430 (S43000)	16-18 Cr, 0.12 C, 1.0 Mn, 1.0 Si, 0.040 P, 030 S, 0.75 Ni	442 (S44200)	18-23 Cr, 0.20 C, 1.0 Mn, 1.0 Si, 0.040 P, 0.030 S
430F (S43020)	16-18 Cr, 0.12 C, 1.25 Mn, 1.0 Si, 0.060 P, 0.15 S min, 0.60 Mo (optional)	446 (S44600)	23-27 Cr, 0.20 C, 1.5 Mn, 1.0 Si, 0.040 P, 0.030 S, 0.025 N

30

continued on next page

Martensitic			
403 (S40300)	11.5-13.0 Cr, 1.15 C, 1.0 Mn, 0.5 Si, 0.040 P, 0.030 S, 0.60 Ni	420F (S42020)	12-14 Cr, over 0.15 C, 1.25 Mn, 1.0 Si, 0.060 P, 0.15 S min, 0.60 Mo max (optional)
410 (S41000)	11.5-13.5 Cr, 0.15 C, 1.0 Mn, 1.0 Si, 0.040 P, 0.030 S, 0.75 Ni	422 (S42200)	11-12.50 Cr, 0.50-1.0 Ni, 0.20- 0.25 C, 0.50-1.0 Mn, 0.50 Si, 0.025 P, 0.025 S, 0.90-1.25 Mo, 0.20-0.30 V, 0.90-1.25 W
414 (S41400)	11.5-13.5 Cr, 1.25-2.50 Ni, 0.15 C, 1.0 Mn, 1.0 Si, 0.040 P, 0.030 S, 1.25-2.50 Ni	431 (S41623)	15-17 Cr, 1.25-2.50 Ni, 0.20 C, 1.0 Mn, 1.0 Si, 0.040 P, 0.030 S
416 (S41600)	12-14 Cr, 0.15 C, 1.25 Mn, 1.0 Si, 0.060 P, 0.15 S min, 0.060 Mo (optional)	440A (S44002)	16-18 Cr, 0.60-0.75 C, 1.0 Mn, 1.0 Si, 0.040 P, 0.030 S, 0.75 Mo
416Se (S41623)	12-14 Cr, 0.15 C, 1.25 Mn, 1.0 Si, 0.060 P, 0.060 S, 0.15 Se min	440B (S44003)	16-18 Cr, 0.75-0.95 C, 1.0 Mn, 1.0 Si, 0.040 P, 0.030 S, 0.75 Mo
420 (S42000)	12-14 Cr, 0.15 C min, 1.0 Mn, 1.0 Si, 0.040 P, 0.030 S	440C (S44004)	16-18 Cr, 0.95-1.20 C, 1.0 Mn, 1.0 Si, 0040 P, 0.030 S, 0.75 Mo

Martensitic: Contains about 1% carbon. Can be heat treated and hardened. Magnetic.

Heat-Resisting			
501 (S50100)	4-6 Cr, 0.10 C min, 1.0 Mn, 1.0 Si, 0.040 P, 0.030 S, 0.40-0.65 Mo	502 (S50200)	4-6 Cr, 0.10 C. 1.0 Mn, 1.0 Si, 0.040 P, 0.030 S, 0.40-0.65 Mo

Heat-Resisting: Contains about 1% carbon. Used for high heat applications. Magnetic.

Machining Stainless Steels

Machining stainless steels is more difficult than machining mild steel, but it need not be problematic. Here are some suggestions and techniques for machining stainless steels:

- All stainless steels machine better when slightly hard.
- Spindle speeds (RPM) of the machine tool are about half that of mild steel.
- Use flood coolant.
- Do not allow the cutter to dwell during cutting.
- If cutter breakdown is excessive, try *increasing* the feed rate.

Clamp the workpiece securely!

Navigation Hint: For more information on machining stainless steels, see Section 6, Unit B, Machining Operations, *Machinery's Handbook Made* Easy; pages 123–140

Unit E: Hardening, Tempering, and Annealing pages 461–512

Hardening

The properties of metal can be changed by controllably heating and cooling the material. *Hardening*, *tempering*, and *annealing* can be accomplished by a process of heating and cooling. Heat treating usually occurs in an oven and the changes that occur happen at the molecular level. The heat treating technician follows a recipe much like a cook bakes a cake. Ingredients must be in the correct proportions, the oven is set at a specific temperature, and the object is baked for a specific amount of time, and cooled. Another similarity between heat treating and baking is that the molecular structure of the object that comes out of the oven is different than when it went in. It is literally a different material.

Quenching occurs immediately after the part comes out of the oven. Quenching mediums vary depending on the material. Because the cooling process is so critical, different quenching mediums are used to control the time it takes to cool the material. Quenching mediums include:

- Water
- Water with salt (brine)
- Oil
- Air
- Molten salt

faster

slower

Steels with high alloy content require slower quenching than steels with fewer alloys. Slower quenching results in deeper hardness. Fast quenching methods such as water produce a thin case of hardness. The process of quenching can be described as a *Time-Temperature-Transformation* development:

The **Time** it takes to quench the material–

Changes the **Temperature** in a controlled fashion so that–

Transformation of the molecular structure is trapped in the desirable arrangement.

Parts that are quenched too fast or too slow will not have the correct properties.

Tempering

Tempering follows hardening and quenching. The material is reheated to a temperature lower than the hardening temperature to reduce internal stresses that may cause the material to crack or otherwise fail in service. Tempering also reduces the hardness of the material and increases toughness and impact resistance.

Application for tempering: Virtually all hardened steel is tempered to prevent it from cracking or breaking.

Annealing

Annealing is the process of softening a metal. The material is placed in an oven or furnace and heated for a specific period of time and allowed to cool slowly to room temperature.

Application for annealing: A feature of a steel part was mistakenly left out and the part was hardened. The part is not machinable by common methods so it is annealed; the machining is completed and the part is re-hardened.

Stress Relieving

This process reduces internal stresses commonly caused by machining. The material is heated to a temperature below the materials' critical range and held there until the temperature evens out.

Properties, Treatment, and Testing of Materials

Application for stress relieving: A large welded steel fixture base is machined and it is inspected before finishing. It is determined that the material is twisting and warping. The weldment is stress relieved and the material remains stable during the finish machining.

Unit F: Nonferrous Alloys pages 513–550

Nonferrous metal is metal that does not contain iron. A nonferrous alloy is a metal that is made up of two or more nonferrous metals. Examples of nonferrous alloys include:

- Brass
- Bronze
- Aluminum Alloys

Shop Recommended pages 515-518, Table 2

CAST COPPER ALLOYS   515

Table 2. Properties and Applications of Cast Coppers and Copper Alloys

UNS Designation	Nominal Composition (%)	Typical Mechanical Properties, as Cast or Heat Treated[a]				Typical Applications
		Tensile Strength (ksi)	Yield Strength (ksi)	Elonga-tion in 2 in. (%)	Machin-ability Rating[b]	
		Copper Alloys				
C80100	99.95 Cu + Ag min, 0.05 others max	25	9	40	10	Electrical and thermal conductors; corrosion and oxidation-resistant applications.
C80300	99.95 Cu + Ag min, 0.034 Ag min, 0.05 others max	25	9	40	10	Electrical and thermal conductors; corrosion and oxidation-resistant applications.
C80500	99.75 Cu + Ag min, 0.034 Ag min, 0.02 B max, 0.23 others max	25	9	40	10	Electrical and thermal conductors; corrosion and oxidation-resistant applications.
C80700	99.75 Cu + Ag min, 0.02 B max, 0.23 others max	25	9	40	10	Electrical and thermal conductors; corrosion and oxidation-resistant applications.
C80900	99.70 Cu + Ag min, 0.034 Ag min, 0.30 others max	25	9	40	10	Electrical and thermal conductors; corrosion and oxidation-resistant applications.
C81100	99.70 Cu + Ag min, 0.30 others max	25	9	40	10	Electrical and thermal conductors; corrosion and oxidation resstant applications.

Copper is an element present in many nonferrous alloys. Table 2 shows the UNS Designation, which is an alloy designation system. The "Nominal Composition" is the percentage of alloying elements.

3

Shop Recommended page 514

Classification of Copper and Copper Alloys

Family	Principal Alloying Element	UNS Numbers[a]
Coppers, high-copper alloys		C1xxxx
Brasses	Zn	C2xxxx, C3xxxx, C4xxxx, C66400 to C69800
Phosphor bronzes	Sn	C5xxxx
Aluminum bronzes	Al	C60600 to C64200
Silicon bronzes	Si	C64700 to C66100
Copper nickels, nickel silvers	Ni	C7xxxx

[a] Wrought alloys.

A prefix of "C" indicates copper alloys, including brass and bronze alloys.

Aluminum Alloys

Aluminum is an element. In its natural state it is very soft and malleable. When alloyed with other metals, desirable properties are obtained such as strength, machineability, thermal conductivity, and corrosion resistance. Aluminum is non-magnetic and non-sparking, and it has excellent conductivity. It can be cast by any method known.

Navigation Hint: For more information on machining aluminum, see Section 6, Unit B Machining Operations; in *Machinery's Handbook Made* Easy, pages 123–140.

— Unit A: Cutting Speeds and Feeds, pages 1008–1020
— Unit B: Speed and Feed Tables, pages 1021–1080

Index navigation path and key words:

— aluminum/machining/page 1192

Nickel and Nickel Alloys

Nickel is a white metal with good corrosion resistance. Nickel and Nickel alloys are used in applications when high strength at high temperature are required. Typical uses of nickel alloys include food processing equipment, springs, turbine and furnace parts, and heat treating equipment.

Navigation Hint: For more information on machining nickel alloys, see Section 6, Unit B Machining Operations; in *Machinery's Handbook Made* Easy, pages 123–140.

— Unit A: Cutting Speeds and Feeds, pages 1008–1020
— Unit B: Speed and Feed Tables, pages 1021–1080

Index navigation path and key words:

— nickel alloys/machining/pages 1125–1126

> *Analyze, Evaluate, & Implement*

3

Use the Machinery's Handbook to identify six types of steel that contain nickel as the main alloying element, and give an application for each.

Titanium and Titanium Alloys

Titanium has a better strength-to-weight ratio of any other metal. It is lighter than aluminum, corrosion resistant, non magnetic, and acid resistant. Titanium and titanium alloys are used extensively in the aircraft industry.

> *Analyze, Evaluate, & Implement*

Learning activity for a group: Across the top of a whiteboard, write down these headings:

- *Carbon Steels*
- *Alloy Steels*
- *Super-Alloy Steels*
- *Stainless Steel*
- *Cast Iron*
- *Ferrous Alloys*
- *Non-Ferrous Alloys*
- *Precious Metals*
- *Tool Steel*
- *Non-Metals*
- *Other*

On separate index cards or sticky notes, write down as many manufacturing materials that you can think of. Concentrate on metals (metal is any material that reflects light and conducts electricity).

> *Place the names of the materials under the proper category.*

Unit G: **Plastics page 551–608**

Shop Recommended page 551: Properties of Plastics

PLASTICS

Properties of Plastics

Characteristics of Important Plastics Families

ABS (acrylonitrile-butadiene-styrene)	Rigid, low-cost thermoplastic, easily machined and thermo-formed.
Acetal	Engineering thermoplastic with good strength, wear resistance, and dimensional stability. More dimensionally stable than nylon under wet and humid conditions.
Acrylic	Clear, transparent, strong, break-resistant thermoplastic with excellent chemical resistance and weatherability.
CPVC (chlorinated PVC)	Thermoplastic with properties similar to PVC, but operates to a 40-60°F (14-16°C) higher temperature.
Fiberglass	Thermosetting composite with high strength-to-weight ratio, excellent dielectric properties, and unaffected by corrosion.
Nylon	Thermoplastic with excellent impact resistance, ideal for wear applications such as bearings and gears, self-lubricating under some circumstances.
PEEK (polyetheretherketone)	Engineering thermoplastic, excellent temperature resistance, suitable for continuous use above 500°F (260°C), excellent flexural and tensile properties.
PET (polyethyleneterephthalate)	Dimensionally stable thermoplastic with superior machining characteristics compared to acetal.
Phenolic	Thermosetting family of plastics with minimal thermal expansion, high compressive strength, excellent wear and abrasion resistance, and a low coefficient of friction. Used for bearing applications and molded parts.
Polycarbonate	Transparent tough thermoplastic with high impact strength, excellent chemical resistance and electrical properties, and good dimensional stability.
Polypropylene	Good chemical resistance combined with low moisture absorption and excellent electrical properties, retains strength up to 250°F (120°C).
Polysulfone	Durable thermoplastic, good electrical properties, operates at temperatures in excess of 300°F (150°C).
Polyurethane	Thermoplastic, excellent impact and abrasion resistance, resists sunlight and weathering.
PTFE (polytetrafluoroethylene)	Thermoplastic, low coefficient of friction, withstands up to 500°F (260°C), inert to chemicals and solvents, self-lubricating with a low thermal-expansion rate.
PVC (polyvinyl chloride)	Thermoplastic, resists corrosive solutions and gases both acid and alkaline, good stiffness.
PVDF (polyvinylidene-fluoride)	Thermoplastic, outstanding chemical resistance, excellent substitute for PVC or polypropylene. Good mechanical strength and dielectric properties.

Commercial plastics are resins containing additives much like steel contains alloys. There are two main classes of resins: thermoplastics and thermosets.

Thermoplastics soften when they are heated. They can be repeatedly melted by heating and solidified by cooling. Recycling of thermoplastics is easy, but repeated heating, cooling, and recycling reduce mechanical properties and affects appearance. Families of thermoplastics include:

- polystyrenes (PS)
- polyethylenes (PE)
- acrylics (PMMA)
- cellulosics (CAB cellulose acetate butyrate)
- polyvinyls

Thermosets behave differently than thermoplastics because they form chemical bonds when heated. Thermosets have good heat resistance, but when heated above their molding temperatures, they decompose. This type of resin cannot be reprocessed, so recycling is limited.

Thermoplastics may be classified by their structure:

- Amorphous
- Crystalline
- Liquid-Crystalline Polymers (LCP)

Manufacture of Plastic Products

The main manufacturing processes for thermoplastic products are:

- extrusion
- injection molding
- blow molding
- sheet thermoforming

Manufacturing processes for thermosetting products are:

- compression molding
- transfer molding
- prepreg molding
- pultrusion

Check these pages for more information on plastics manufacturing:

— plastics/blow molding/page 581
— plastics/sheet thermoforming/page 581
— plastics/processing thermosets/page 581

Assembly with Fasteners

Examples of poor and preferred designs of plastic assemblies are shown in Figure 26. Special considerations are necessary to allow for the flexibility and expansion of plastic.

3

Shop Recommended page 599, Figure 26

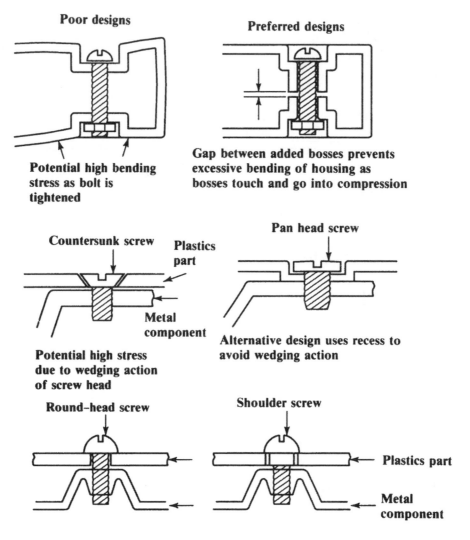

Figure 3.26 Examples of bad and good designs in assembling plastics with metal fasteners

Machining Plastics

Plastics can be molded into complex shapes and usually do not require further finishing operations. However, there are times when machining plastic is desirable:

- low volume of parts do not permit the building of a complex tool
- undercuts or openings that would be hard to mold
- prototype development requires frequent changes

Safety First

- **All machining of plastics requires dust control, adequate ventilation, safety guards, and eye protection.**
- **Materials, including plastics, are required to have a Material Safety Data Sheet (MSDS)**

   A Material Safety Data Sheet (MSDS) is designed to provide both workers and emergency personnel with the proper procedures for handling or working with a particular substance. MSDS's include information such as physical data (melting point, boiling point, flash point etc.), toxicity, health effects, first aid, reactivity, storage, disposal, protective equipment, and spill/leak procedures.

Shop Recommended page 598–600 Machining Plastics and Thermo Properties

Machining Plastics.—Plastics can be molded into complex shapes and so do not usually need to be machined. However, machining is sometimes more cost-effective than making a complex tool, especially when requirements are for prototype development, low-volume production, undercuts, angular holes, or other openings that are difficult to produce in a mold. Special methods for development of prototypes are discussed later. All machining of plastics requires dust control, adequate ventilation, safety guards, and eye protection.

Like some metals, plastics may need to be annealed before machining to avoid warpage. Some commercially available bar and rod stock are sold already annealed. If annealing is necessary, instructions can be obtained from plastics suppliers. Plastics moduli are small fractions—2 to 10 per cent—of those of metals and this lower stiffness permits much greater deflection of the work material during cutting. Thermoplastics materials must be held and supported firmly to prevent distortion, and sharp tools are essential to minimize normal forces.

Turning and Cutting Off

Turning and Cutting Off: High speed steel and carbide tools are commonly used with cutting speeds of 200-500 and 500-800 ft/min (61-152 and 152-244 m/min), respectively. Water-soluble coolants can be used to keep down temperatures at the shear zone and improve the finish, except when they react with the work material. Chatter may result from the low modulus of elasticity and can be reduced by close chucking and follow rests. Box tools are good for long, thin parts. Tools for cutting off plastics require greater front and side clearances than are needed for metal. Cutting speeds should be about half those used for turning operations.

Making it Simple

- Turning and Cutting Off are lathe operations.
- Some types of coolants cause plastic to degrade.
- Hold the plastic workpiece as close to the clamping device as possible.

Drilling

Drilling: This is the most common machining operation because small-diameter holes are more easily drilled than molded. However, plastics are rather difficult to drill without some damage. Many difficulties not encountered in drilling metals, such as gumming, burning in the drilled hole, cracks around the edges or growth of cracks after drilling, can occur. Two reasons for these difficulties are: swarf flow (chip removal) in drilling is poor, and cutting speeds vary from the center to the periphery of the drill, so that drilling imposes severe loading on the workpiece. Some drill types used with plastics are shown in Fig. 27.

Making it Simple

- Drilling plastic is difficult because the material is too soft.
- Drills that have special point styles work better than drills that are made for steel.

The Thermal Expansion of Plastic

Plastics have thermal expansion coefficients some 10 times higher than those of metals so that even though actual heat generation during machining may be less than with metals there can easily be more expansion. Adequate tool clearances must be provided to minimize heating. Compared with most structural metals, temperatures at which plastics soften, deform and degrade are quite low. Allowing frictional heat to build up causes gumming, discoloration, poor tolerance control, and rough finishes. These effects are more pronounced with plastics such as polystyrene and polyvinyl chloride that have low melting points than with plastics that have higher melting points, such as nylons, fluoroplastics, and polyphenylene sulfide. Sufficient clearances must be provided on cutting tools to prevent rubbing contact between the tool and the work. Tool surfaces that will come into contact with plastics during machining should be polished to reduce frictional drag and resulting temperature increases. Proper rake angles depend on depth of cut, cutting speed, and the type of plastic being cut. Large rake angles should be used to produce continuous-type cuttings, but they should not be so large as to cause brittle fracture of the work and resulting discontinuous chips.

Making it Simple

- Plastics expand much more than steel given the same temperature increase.
- Cutting tools should have sharp, polished edges.
- Continuous chips are desirable.

Navigation Hint: For more information on Cutting Speed and Feed Formulas, see Section 6; Machining Operations in *Machinery's Handbook Made Easy,* pages 120–160:

— Unit A Cutting Speeds and Feeds, pages 1008-1020
— Unit B Speed and Feed Tables, pages 1021–1080

3

Shop Recommended page 600 Drill Designs Used for Plastics

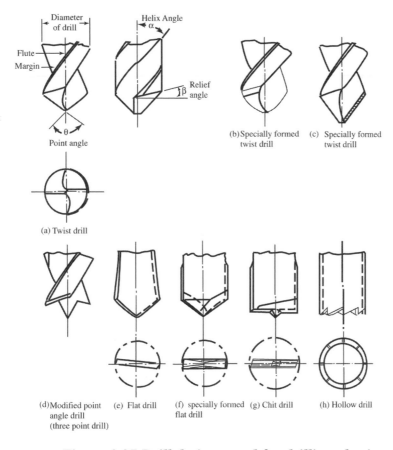

Figure 3.27 Drill designs used for drilling plastics

Shop Recommended page 600

Making it Simple

- The part of the drill that has the spiral channel is where the helix angle is measured.
- Changing the helix angle to a flatter angle helps pull the chips out of the hole.
- As a rule, softer materials are drilled with smaller point angles.
- Harder materials are drilled with flatter point angles.

Shop Recommended page 601, Table 8

Drilling and reaming speed and feed suggestions for various materials are shown in Table 8. These speeds and feeds can be increased where there is no melting, burning, discoloration, or poor surface finish. Drilling is best done with commercially available drills designed for plastics (Fig. 27), usually having large helix angles, narrow lands, and highly polished or chromium-plated flutes to expel swarf rapidly and minimize frictional heating. Circle cutters are often preferred for holes in thin materials. Drills must be kept sharp and cool, and carbide tools may be needed in high production, especially with glass-reinforced materials. They must be cooled with clean compressed air to avoid contamination. Aqueous solutions are used for deep drilling because metal cutting fluids and oils may degrade or attack the plastics and may cause a cleaning problem. Plastics parts must be held firmly during drilling to counter the tendency for the tooling to grab and spin the work.

Table 8. Speeds and Feeds for Drilling Holes of 0.25 to 0.375 inch (6.3–9.5 mm) Diameter in Various Thermoplastics

Material	Speed (rpm)	Feed[a]	Comments
Polyethylene	1,000–2,000	H	Easy to machine
Polyvinyl chloride	1,000–2,000	M	Tends to become gummy
Acrylic	500–1,500	M-H	Easy to drill with lubricant
Polystyrene	500–1,500	H	Must have coolant
ABS	500–1,000	M-H	
Polytetrafluoroethylene	1,000	L-M	Easy to drill
Nylon 6/6	1,000	H	Easy to drill
Polycarbonate	500–1,500	M-H	Easy to drill, some gumming
Acetal	1,000–2,000	H	Easy to drill
Polypropylene	1,000–2,000	H	Easy to drill
Polyester	1,000–1,500	H	Easy to drill

[a] H = high; M = medium; L = low.

Making it Simple

- If plastic starts to melt while machining, reduce spindle speed.
- Drills for thin plastic are shaped like tubes with cutting teeth on the flat end.
- Plastic with fiberglass in it is very abrasive and wears out cutters easily.

Shop Recommended page 600–601

Tapping and Threading of Plastics: Many different threaded fasteners can be used with plastics, including thread-tapping and -forming screws, threaded metal inserts, and molded-in threads, but threads must sometimes be machined after molding. For tapping of through-holes in thin cast, molded, or extruded thermoplastics and thermosets, a speed of 50 ft/min (15.2 m/min) is appropriate. Tapping of filled materials is done at 25 ft/min (7.6 m/min). These speeds should be reduced for deep or blind holes and when the percentage of thread is greater than 65-75 per cent. Taps should be of M10, M7, or M1, molybdenum high-speed steel, with finish-ground and -polished flutes. Two-flute taps are recommended for holes up to 0.125 inch (3.2 mm) diameter. Oversize taps may be required to make up for elastic recovery of the plastics. The danger of retapping on the return stroke can be reduced by blunting the withdrawal edges of the tool.

Making it Simple

- Special taps for cutting threads in plastic have polished flutes.
- Oversize taps are available to make up for shrinkage after tapping.
- The percentage of thread is determined by the size of the tap drill.

Shop Recommended page 2030

2030 TAPPING

Table 4. Tap Drills and Clearance Drills for Machine Screws with American National Thread Form

Size of Screw		No. of Threads per Inch	Tap Drills		Clearance Hole Drills			
No. or Diam.	Decimal Equiv.		Drill Size	Decimal Equiv.	Close Fit		Free Fit	
					Drill Size	Decimal Equiv.	Drill Size	Decimal Equiv.
0	.060	80	3/64	.0469	52	.0635	50	.0700
1	.073	64	53	.0595	48	.0760	46	.0810
		72	53	.0595				
2	.086	56	50	.0700	43	.0890	41	.0960
		64	50	.0700				
3	.099	48	47	.0785	37	.1040	35	.1100
		56	45	.0820				
4	.112	36[a]	44	.0860	32	.1160	30	.1285
		40	43	.0890				
		48	42	.0935				
5	.125	40	38	.1015	30	.1285	29	.1360
		44	37	1040				
6	.138	32	36	.1065	27	.1440	25	.1495
		40	33	.1130				
8	.164	32	29	.1360	18	.1695	16	.1770
		36	29	.1360				
10	.190	24	25	.1495	9	.1960	7	.2010
		32	21	.1590				
12	.216	24	16	.1770	2	.2210	1	.2280
		28	14	.1820				
14	.242	20[a]	10	.1935	D	.2460	F	.2570
		24[a]	7	.2010				
1/4	.250	20	7	.2010	F	.2570	H	.2660
		28	3	.2130				
5/16	.3125	18	F	.2570	P	.3230	Q	.3320
		24	I	.2720				
3/8	.375	16	5/16	.3125	W	.3860	X	.3970
		24	Q	.3320				
7/16	.4375	14	U	.3680	29/64	.4531	15/32	.4687
		20	25/64	.3906				
1/2	.500	13	27/64	.4219	31/64	.5156	17/32	.5312
		20	29/64	.4531				

[a] These screws are not in the American Standard but are from the former A.S.M.E. Standard.

The size of the tap drill hole for any desired percentage of full thread depth can be calculated by the formulas below. In these formulas the Per Cent Full Thread is expressed as a decimal; e.g., 75 per cent is expressed as .75. The tap drill size is the size nearest to the calculated hole size.

For American Unified Thread form:

$$\text{Hole Size} = \text{Basic Major Diameter} - \frac{1.08253 \times \text{Per Cent Full Thread}}{\text{Number of Threads per Inch}}$$

SECTION 3

Analyze, Evaluate, & Implement

Calculate the tap drill sizes for 3/8-16 and 3/4-10 taps

- *for 60% thread*
- *for 75% thread*

Shop Recommended page 602, Table 9

Sawing Thermoset Cast or Molded Plastics: Circular or band saws may be used for sawing. Circular saws provide smoother cut faces than band saws, but band saws run cooler so are often preferred even for straight cuts. Projection of the circular saw above the table should be minimized. Saws should have skip teeth or buttress teeth with zero front rake and a raker set. Precision-tooth saw blades should be used for thicknesses up to 1 inch (25.4 mm), and saws with buttress teeth are recommended for thicknesses above 1 inch (25.4 mm). Dull edges to the teeth cause chipping of the plastics and may cause breakage of the saw. Sawing speeds and other recommendations for using blades of high-carbon steel are shown in Table 9.

Table 9. Speeds and Numbers of Teeth for Sawing Plastics Materials with High-Carbon Steel Saw Blades

Material Thickness		Number of Teeth on Blade	Peripheral Speed			
			Thermoset Cast or Molded Plastics		Thermoplastics (and Epoxy, Melamine, Phenolic and Allyl Thermosets)	
(inch)	(mm)		(ft/min)	(m/min)	(ft/min)	(m/min)
0–0.5	0–13	8-14	2000–3000	607–914	4000–5000	1219–1524
0.5–1	13–25	6-8	1800–2200	549–671	3500–4300	1067–1311
1–3	25–76	3	1500–2200	475–671	3000–3500	914–1067
>3	>76	>3	1200–1800	366–549	2500–3000	762–914

Making it Simple

- ■ Saw blades are available with several different tooth styles.
- ■ Saw blades cut a path wider than the blade thickness because the cutting teeth are staggered.

Shop Recommended page 602

Milling of Plastics: Peripheral cutting with end mills is used for edge preparation, slotting and similar milling operations, and end cutting can also be used for facing operations. Speeds for milling range from 800 to 1400 ft/min (244–427 m/min) for peripheral end milling of many thermoplastics and from 400 to 800 ft/min (122–244 m/min) for many thermosets. However, slower speeds are generally used for other milling operations, with some thermoplastics being machined at 300-500 ft/min (91–152 m/min), and some thermosets at 150-300 ft/min (46–91 m/min). Adequate support and suitable feed rates are very important. A table feed that is too low will generate excessive heat and cause surface cracks, loss of dimensional accuracy, and poor surface finish. Too high a feed rate will produce a rough surface. High-speed steel tools (M2, M3, M7, or T15) are generally used, but for glass-reinforced nylon, silicone, polyimide, and allyl, carbide (C2) is recommended.

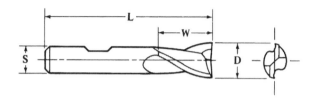

Figure 3.1 End mill

Making it Simple

■ End mills cut on both the periphery and on the bottom.
■ The bottom of the end mill is used for "facing operations."
■ Use the fastest feed rate that gives the required surface finish.

Navigation Hint: For more information on High Speed Steel and Carbide Cutting Tools, see *Machinery's Handbook Made* Easy, Section 6, Machining Operations, pages 123–140

— Unit A: Cutting Speeds and Feeds
— Unit B: Speed and Feed Tables

Shop Recommended page 601–602

Other Machining Techniques: Lasers can be used for machining plastics, especially sheet laminates, although their use may generate internal stresses. Ultrasonic machining has no thermal, chemical, or electrical reaction with the workpiece and can produce holes down to 0.003-inch (0.0762 mm) diameter; tight tolerances, 0.0005 inch (0.0127 mm); and very smooth finishes, 0.15 μinch (0.381 μm) with No. 600 boron carbide abrasive powder. Water-jet cutting using pressures up to 60,000 lb/inch^2 (414 N/mm^2) is widely used for plastics and does not introduce stresses into the material. Tolerances of ± 0.004 inch (± 0.102 mm) can be held, depending on the equipment available. Process variables, pressures, feed rates, and the nozzle diameter depend on the material being cut. This method does not work with hollow parts unless they can be filled with a solid core.

SECTION 3

Making it Simple

- The word LASER is an acronym for Light Amplification by the Stimulated Emission of Radiation.
- Ultrasonic Machining uses high frequency vibrations.
- Water jet cutting machines are preferred for machining plastic.

3

Unit A: Elements, Heat, Mass, and Weight pages 371–384

 Index navigation paths and key words:

Unit B: Properties of Wood, Ceramics, Plastics, Metals pages 385-395

 Index navigation paths and key words:

ASSIGNMENT

List the key terms and give a definition of each.

Alloy	Harden
Element	Anneal
Hardened	Stress relieve
Tempered	Thermosets
Pig Iron	Thermoplastics

Section 4

DIMENSIONING, GAGING, and MEASURING

Use this section with pages 609 — 753
in Machinery's Handbook 29th Edition

SECTION 4

Navigation Overview

Units Covered in this Section

Navigation Assistant
Watch for Navigation Hints

The Navigation Assistant helps find information in the MH29 Primary Index. The Primary Index is located in the back of the book on pages 2701-2788 and is set up alphabetically by subject. Watch for the magnifying glass throughout the section for navigation hints.

Key Terms

Meter	Basic Dimension
Millimeter	Interference Fit
Inch	Tolerance
Force Fit	Discrimination
Visible Line	Maximum Material
Hidden Line	Condition
Datum	

Learning Objectives

After studying this unit you should be able to:

- Convert millimeters to inches and inches to millimeters.
- Create a sketch using proper drawing practices.
- Describe three methods of assembling an interference fit.
- Read a 0-1″ micrometer within a ten-thousandth of an inch.
- Read a vernier scale within a thousandth of an inch.
- Interpret surface finish symbols.
- State three advantages of Geometric Dimensioning and Tolerancing (GD&T).
- Interpret a typical GD&T feature control frame.

4

Introduction

Section 4 of MH 28 relates to some of the most basic machine shop practices. For example, the assembly of machines, tools, and dies is not possible without a plan or the ability to measure components. The proper fit between mating parts is often critical. The surface texture (or smoothness) of parts made to a high degree of accuracy is directly related to the degree of accuracy required.

Unit A: Drafting Practices pages 609–627

Index navigation paths and key words:

— drafting practices/ANSI standard/page 609

Unit A, "Drafting Practices" has its roots in a time when engineering drawings were hand drawn. The process of making the shop copy of the drawing generated a document that was actually blue in color, hence the term "blueprint." This is not done anymore, but the ability to create detailed pencil sketches is an invaluable skill. Communication between shifts or across departments often includes a sketch to express ideas, plans, and designs. The sketch is not usually an official component of the job in that it is not retained in a formal manner or sent to the customer.

Your proficiency in sketching and attention to detail demonstrates a degree of professionalism. The importance of this cannot be overstated. Table 2 on page 610 identifies the proper way to illustrate the lines used in engineering drawings and sketches.

SECTION 4

Shop Recommended page 610, Table 2

DRAFTING PRACTICES

Table 2. American National Standard for Engineering Drawings
ANSI/ASME Y14.2M-1992 (R2008)

4

Visible Lines are also called "object lines" and are the thickest lines on the drawing.

Hidden Lines show interior detail of the part.

Center Lines are used to show the axis of circular features.

Arrowheads touch extension Lines.

Extension Lines do not touch object lines. Leave a space of about 1/16″.

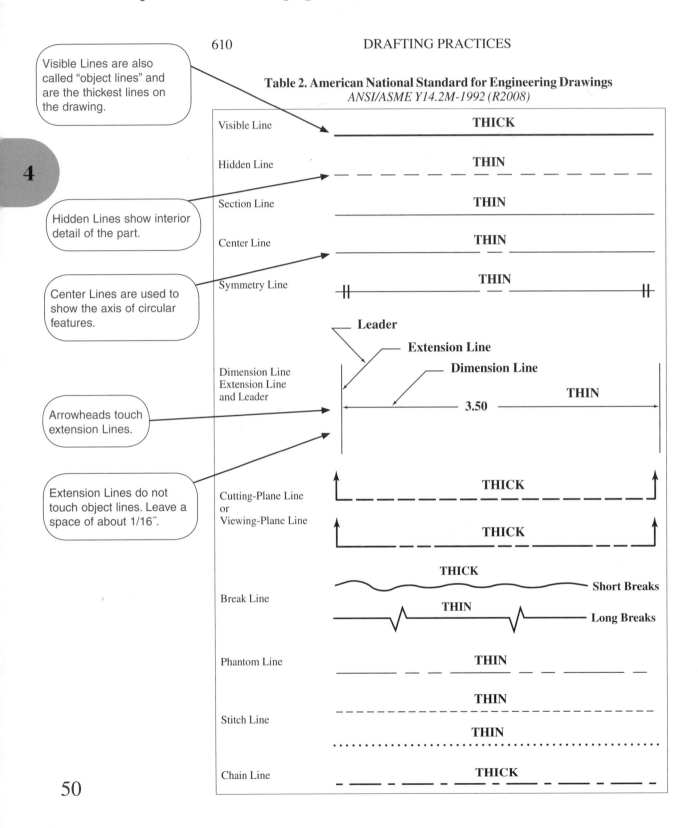

50

Dimensioning, Gaging, and Measuring

See Figure 4.1 for examples of how these lines are used on an engineering drawing. For professional-looking results, use these rules when creating a pencil sketch:

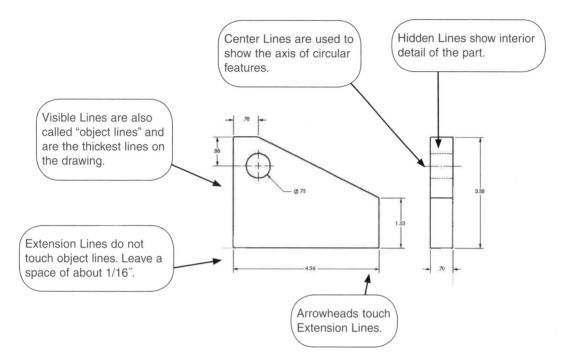

Figure 4.1 Sample engineering drawing

Abbreviations Used on Engineering Drawings

To save space on drawings, it is a common practice to abbreviate terms (sometimes to a fault). Because there is no real standard, communication problems between designers and toolmakers can result from using abbreviations that are not recognizable. The following is a list of many, but not all, abbreviations found on drawings.

Across Flats	ACR FLTS
Assembly	ASSY
Bill of Material	BOM
Bolt Circle	BC
Carbon Steel (Mild Steel, Low Carbon Steel, Cold Rolled Steel)	CS, MS, CRS
Cast Iron	CI
Chamfer	CHAM or CHMF
Counterbore	C BORE
Countersink	C SINK
Deep or Depth	DP
Diameter	DIA
Dimension	DIM

Drawing	DWG
Equally Spaced	EQ SP or EQL SP
Far Side, Near Side	FS, NS
Finish All Over	FAO
Heat Treat	HT
Hexagon	HEX
Inside Diameter, Outside Diameter	ID, OD
International Organization for Standardization	ISO
Left Hand, Right Hand	LH, RH
Long	LG
Material	MTL
Maximum Material Condition	MMC
Metric	M
Minimum	MIN
Near Side, Far Side	NS, FS
Nominal	NOM
Not to Scale	NTS (Dimension may be underlined)
Number	NO
Outside Diameter, Inside Diameter	OD, ID
Oversize	OS
Perpendicular	PERP
Pitch Diameter	PD
Pitch	P
Radius	R
Reference Dimension	(value is inside of parentheses)
Revolutions per Minute	RPM
Right Hand, Left Hand	RH, LH
Section	SEC or SECT
Spherical Radius	SR
Spotface	SP
Square	SQ
Steel	STL
Symmetrical	SYM
Taper Pipe Thread	NPT
Thick	THK
Thread	THD
To Sharp Corner	TSC
Through	THRU
Undercut	UCUT

4

Dimensioning, Gaging, and Measuring

Geometric Dimensioning and Tolerancing pages 613–622

 Index navigation paths and key words:

— geometric dimensioning and tolerancing/page 613

Geometric Dimensioning and Tolerancing (GD&T) is a system that uses symbols to convey specifications on engineering drawings and related documents. The use of GD&T ensures continuity between the designer, the machinist, and the inspector because everyone involved in the manufacturing process will use the same *datums* (point from which dimensions are taken). Another advantage of GD&T is that it guarantees proper fit and interchangeability between mating components.

Shop Recommended pages 613 and 618, Figure 10b

"Geometric dimensioning and tolerancing provides a comprehensive system for symbolically defining the geometric tolerance zone within which features must be contained."

For an illustration regarding the statement on page 613 shown above, see Fig. 10b.; page 618. This is a typical engineering drawing with GD&T controls and definitions

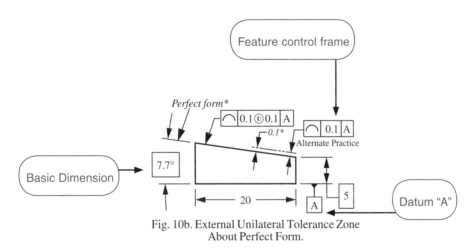

Fig. 10b. External Unilateral Tolerance Zone
About Perfect Form.

Basic Dimension

The dimensions inside of boxes are known as "basic dimensions." A basic dimension is a numerical value used to describe the theoretical exact size of the part feature referenced.

Datum "A"

A datum is a feature that is used as a starting point or a place from which dimensions begin. GD&T controls such as parallelism and perpendicularity are referenced to a datum feature. A datum can be a point, line, centerline of a hole, or a surface. The designer chooses datums based on the function of the part or assembly.

Feature control frame

A feature control frame is a specification that shows the type of GD&T controls being used to control the feature referenced. Feature control frames can be read like a story:

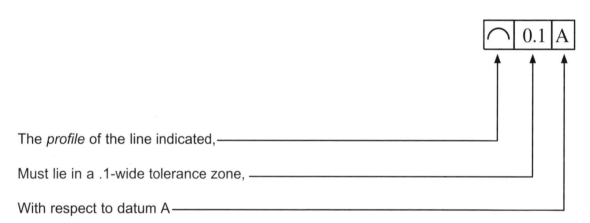

The *profile* of the line indicated,

Must lie in a .1-wide tolerance zone,

With respect to datum A

Dimensioning, Gaging, and Measuring

Shop Recommended page 614, Table 5

The symbols (Geometric Characteristic) used in GD&T

Table 5. ASME Y14.5 Geometric Control Symbols

Type [a]	Geometric Characteristics		Pertains To	Basic Dimensions	Feature Modifier	Datum Modifier
Form	▬	Straightness	Only individual feature	Not applicable	See Table Note 1	No datum
	○	Circularity				
	▱	Flatness			Modifier not applicable	
	⌭	Cylindricity				
Profile	⌒	Profile (Line)	Individual or related	Yes if related		See Table Note 1
	⌓	Profile (Surface)				
Orientation	∠	Angularity	Always related feature(s)	Yes	See Table Note 1	
	⊥	Perpendicularity		Not applicable		
	//	Parallelism				
Location	⊕	Position		Yes		
	◎	Concentricity		Not applicable	Only RFS	
	≡	Symmetry				
Runout	↗	Circular Runout				
	↗↗	Total Runout[b]				

The symbols used in GD&T

The symbols for these controls resemble the characteristics of the area being controlled. For example, Straightness is represented by a straight line. The control Perpendicularity is a line perpendicular (90° to) another line. Parallelism is shown as two parallel lines.

Making It Simple

There are five types of GD&T controls:

■ Form
■ Profile
■ Orientation
■ Location
■ Runout

55

SECTION 4

Form

Straightness, circularity, flatness, and cylindricity controls apply to individual features that do not have a size, such as a surface, an edge, or a centerline (see Figure 4.2).

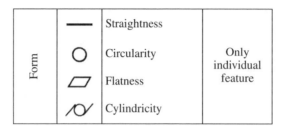

Figure 4.2 Form tolerances

Profile

Profile tolerances include Profile of a Line, and Profile of a Surface (see Figure 4.3). Profile tolerances are similar to form tolerances except they are usually applied to surfaces or lines that are not straight. Profile tolerances may be used with a datum feature.

Figure 4.3 Profile tolerances

Orientation

Orientation tolerances are Angularity, Perpendicularity, and Parallelism (see Figure 4.4). Orientation tolerances must be referenced to a datum feature or features because if a perpendicularity control is used, for example, the machinist needs to know what two features are perpendicular. One surface or edge must be a datum.

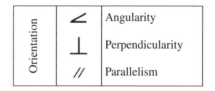

Figure 4.4 Orientation tolerances

Location

Location tolerances are Position, Concentricity, and Symmetry (see Figure 4.5). Location tolerances must be referenced to a datum feature or features.

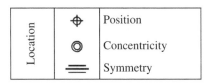

Figure 4.5 Location tolerances

Runout

Runout tolerances are Circular Runout and Total Runout (see Figure 4.6). Runout tolerances must be referenced to a feature or features. A common use of Runout is to control the circular elements of a shaft. If a shaft is perfectly straight, all circular surface elements are in perfect form. Runout is typically checked with a dial indicator while rotating the shaft. The symbol for Runout is easy to remember because the arrow looks like a dial indicator needle.

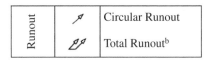

Figure 4.6 Runout

More examples of Feature Control Frames

GD&T controls are shown in a "feature control frame" which can be read like a story:

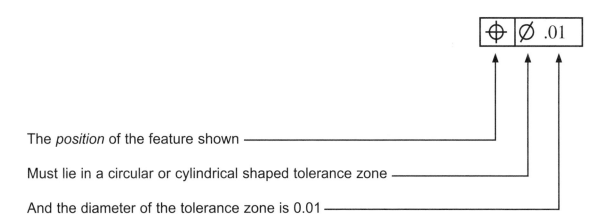

The *position* of the feature shown ——————————

Must lie in a circular or cylindrical shaped tolerance zone ——————

And the diameter of the tolerance zone is 0.01 ——————

57

SECTION 4

Shop Recommended page 616, Basic Dimensions and Definitions

Dimension, Basic: A numerical value used to describe the theoretically exact size, orientation, location, or optionally, the profile, of a feature or datum or datum target. Basic dimensions are indicated by a rectangle around the dimension and are not toleranced directly or by default, see Fig. 6. The specific dimensional limits are determined by the permissible variations as established by the tolerance zone specified in the feature control frame. A dimension is only considered basic for the geometric control to which it is related.

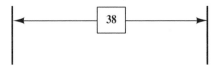

Fig. 6. Basic Dimensions

Dimension, Origin: A symbol used to indicate the origin and direction of a dimension between two features. The dimension originates from the symbol with the dimension tolerance zone being applied at the other feature, see Fig. 7.

Dimension, Reference: A dimension, usually without tolerance, used for information purposes only. Considered to be auxiliary information and not governing production or inspection operations. A reference dimension is a repeat of a dimension or is derived from a calculation or combination of other values shown on the drawing or on related drawings.

Feature Control Frame [Tolerance Frame]: Specification on a drawing that indicates the type of geometric control for the feature, the tolerance for the control, and the related datums, if applicable, see Fig. 8.

Feature: The general term applied to a physical portion of a part, such as a surface, hole, pin, tab, or slot. In ISO practice, depending on how the tolerance frame leader line is attached to the feature, different interpretations may be invoked as to whether the reference is to a line or surface, or an axis or median planer.

58

When an ISO tolerance frame leader line terminates on the outline of the feature, it indicates that the control is a line or the surface itself (Fig. 9a.) When an ISO tolerance frame leader line terminates on a dimension, the axis or medium plane of the dimensioned feature is being controlled. Either inside or outside dimension lines may be used (Fig. 9b.)

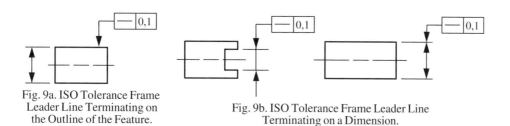

Fig. 9a. ISO Tolerance Frame Leader Line Terminating on the Outline of the Feature.

Fig. 9b. ISO Tolerance Frame Leader Line Terminating on a Dimension.

Feature of Size, Regular: One cylindrical or spherical surface, a circular element, and a set of two opposed parallel elements or opposed parallel surfaces, each of which is associated with a directly toleranced dimension.

Feature of Size, Irregular: A directly toleranced feature or collection of features that may contain or be contained by an actual mating envelope that is: a) a sphere, cylinder, or pair of parallel planes; or, b) other than a sphere, cylinder, or pair of parallel planes.

Least Material Boundary (LMB): The limit defined by a tolerance or combination of tolerances that exist on or inside the material of a feature or features.

Least Material Condition (LMC): The condition in which a feature of size contains the least amount of material within the stated limits of size, for example, upper limit or maximum hole diameter and lower limit or minimum shaft diameter.

Limits, Upper and Lower (UL and LL): The arithmetic values representing the maximum and minimum size allowable for a dimension or tolerance. The upper limit represents the maximum size allowable. The lower limit represents the minimum size allowable.

4

59

SECTION 4

Maximum Material Boundary (MMB): The limit defined by a tolerance or combination of tolerances that exist on or outside the material of a feature or features.

Maximum Material Condition (MMC): The condition in which a feature of size contains the maximum amount of material within the stated limits of size. For example, the lower limit of a hole is the minimum hole diameter. The upper limit of a shaft is the maximum shaft diameter.

Position: Formerly called true position, position is the theoretically exact location of a feature established by basic dimensions. In ISO practice a basic dimension is called a theoretically exact dimension (TED). A positional tolerance is indicated by the position symbol, a tolerance value, applicable material condition modifiers, and appropriate datum references placed in a feature control frame.

Regardless of Feature Size (RFS): The term used to indicate that a geometric tolerance or datum reference applies at any increment of size of the feature within its tolerance limits. RFS is the default condition unless MMC or LMC is specified. The concept is now the default in ASME Y14.5-2009, unless specifically stated otherwise. Thus the symbol for RFS is no longer supported in ASME Y14.5-2009.

Regardless of Material Boundary (RMB) indicates that a datum feature simulator progresses from MMB toward LMB until it makes maximum contact with the extremities of a feature(s). See *Datum Simulator* on page 616.

Size, Actual: The term indicating the size of a feature as produced.

Tolerance Zone Symmetry: In geometric tolerancing, the tolerance value stated in the feature control frame is always a single value. Unless otherwise specified, it is assumed that boundaries created by the stated tolerance are bilateral and equidistant about the perfect form control specified. See Fig. 10a for default zone. If desired, the tolerance may be specified as unilateral or unequally bilateral. See Figs. 10b and 10c for external and internal unilateral zones, and Fig. 10d for an example of a bilateral asymmetrical zone.

Tolerance Zone Symmetry Examples

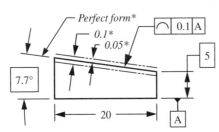

Fig. 10a. Default Symmetrical Tolerance Zone About Perfect Form.

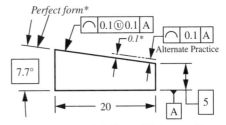

Fig. 10b. External Unilateral Tolerance Zone About Perfect Form.

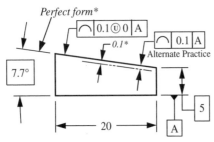

Fig. 10c. Internal Unilateral Tolerance Zone About Perfect Form.

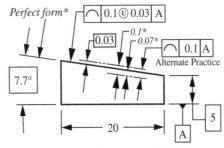

Fig. 10d. Bilateral Asymmetrical Tolerance Zone About Perfect Form.

* Added for clarification and is not part of the specification.

60

Tolerance, Bilateral: A tolerance where variation is permitted in both directions from the specified dimension. Bilateral tolerances may be equal or unequal.

Tolerance, Geometric: The general term applied to the category of tolerances used to control form, profile, orientation, location, and runout.

Tolerance, Unilateral: A tolerance where variation is permitted in only one direction from the specified dimension.

True Geometric Counterpart: Theoretically perfect plane of a specified datum feature.

Virtual Condition: A constant boundary generated by the collective effects of the feature size, its specified MMC or LMC material condition, and the geometric tolerance for that condition.

Shop Recommended page 619–620 Datum Targets

4

Datum Targets: Datum targets are used to establish a datum plane. They may be points, lines or surface areas. Datum targets are used when the datum feature contains irregularities, other features block the surface or the entire surface cannot be used. Examples where datum targets may be indicated include uneven surfaces, forgings and castings, weldments, non-planar surfaces or surfaces subject to warping or distortion. The datum target symbol is located outside the part outline with a leader directed to the target point, area or line. The targets are dimensionally located on the part using basic or toleranced dimensions. If basic dimensions are used, established tooling or gaging tolerances apply.

A solid leader line from the symbol to the target is used for visible or near side locations with a dashed leader line used for hidden or far side locations. The datum target symbol is divided horizontally into two halves. The top half contains the target point area if applicable; the bottom half contains a datum feature identifying letter and target number. Target numbers indicate the quantity required to define a primary, secondary, or tertiary datum. If indicating a target point or target line, the top half is left blank. Datum targets and datum features may be combined to form the datum reference frame, see Fig. 11.

Use of Datum Targets

A common use of datum targets is on castings which have not been machined. The designer will select areas of significance on the part and establish these areas as *datum targets.* These datum targets are used to set up the part for machining. In other words, the part is supported on the targeted areas. As machined surfaces are created, the preferred practice is to use the machined surfaces as new datums.

See page 620 of *Machinery's Handbook 29* for an example of how datum targets are applied (Figure 11) as well as further discussion of datum target lines and datum target areas.

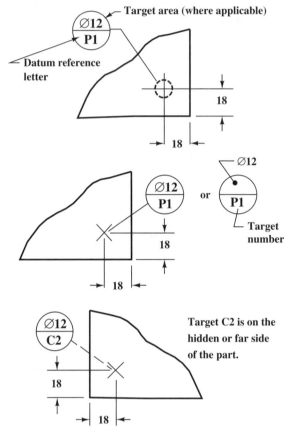

Fig. 11. Datum Target Symbols.

Datum Target Points: A datum target point is indicated by the symbol "X," which is dimensionally located on a direct view of the surface. Where there is no direct view, the point location is dimensioned on multiple views.

Datum Target Lines: A datum target line is dimensionally located on an edge view of the surface using a phantom line on the direct view. Where there is no direct view, the location is dimensioned on multiple views. Where the length of the datum target line must be controlled, its length and location are dimensioned.

Datum Target Areas: Where it is determined that an area, or areas, of flat contact are necessary to ensure establishment of the datum, and where spherical or pointed pins would be inadequate, a target area of the desired shape is specified. Examples include the need to span holes, finishing irregularities, or rough surface conditions. The datum target area may be indicated with the "X" symbol as with a datum point, but the area of contact is specified in the upper half of the datum target symbol. Datum target areas may additionally be specified by defining controlling dimensions and drawing the contact area on the feature with section lines inside a phantom outline of the desired shape.

Positional Tolerance.—A positional tolerance defines a zone within which the center, axis, or center plane of a feature of size is permitted to vary from true (theoretically exact) position. Basic dimensions establish the true position from specified datum features and between interrelated features. A positional tolerance is indicated by the position symbol, a tolerance, and appropriate datum references placed in a feature control frame.

Datum features

Index navigation paths and key words:

— geometric dimensioning and tolerancing/datum feature/page 615

Feature control frames can contain references to more than one datum and may have "modifiers." Translate the feature control frame seen in Figure 4.7.

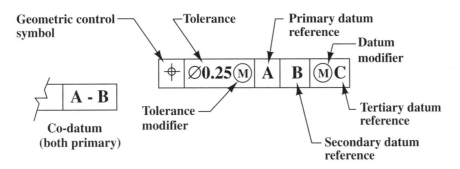

Figure 4.7 Sample feature control frame

Geometric control symbol
This is the characteristic being controlled–in this case, True Position (the position of a feature such as a hole).

Tolerance
The allowable deviation from perfect form.

Primary datum reference
The primary datum is the datum that is referenced first, such as a surface. A surface is defined by three points in space.

Secondary datum reference
The secondary datum is datum referenced second. A secondary datum may be an edge (or line). Lines are defined by two points in space.

Tertiary datum reference
The tertiary datum is the datum referenced third. A tertiary datum may be a point. A point is defined as one point in space.

Tolerance modifier

The tolerance modifier in this feature control frame is an "M," meaning that the 0.25 tolerance applies when the feature is at its *maximum material condition* or MMC. A hole is at its MMC when it is the smallest because that is when it contains the most material. A shaft is at its MMC when it is at its largest size. As the feature being referenced departs from its MMC, a bonus tolerance is added to the 0.25 tolerance equal to amount of the departure.

Shop Recommended pages 620–621

Positional Tolerance.—A positional tolerance defines a zone within which the center, axis, or center plane of a feature of size is permitted to vary from true (theoretically exact) position. Basic dimensions establish the true position from specified datum features and between interrelated features. A positional tolerance is indicated by the position symbol, a tolerance, and appropriate datum references placed in a feature control frame.

Modifiers: In certain geometric tolerances, modifiers in the form of additional symbols may be used to further refine the level of control.

More on the Position Tolerance and Material Modifiers

A typical use of the *Position Tolerance* symbol is to designate the location of a feature, such as a hole. In Figure 13 on page 622, the position symbol is shown as a "cross hair" followed by a diameter symbol. The diameter symbol means that the shape of the tolerance zone is diametrical or cylindrical. In this case, the size of the zone is 0.25. The centerline of the hole must be inside the 0.25 zone.

The "M" means that a bonus tolerance is available based on the size of the feature. The 0.25 mm tolerance applies when the feature is machined at its *"maximum material condition"* (MMC). Female features such as holes are at their maximum material condition when they are at their smallest size (when they have the most material). External features such as shafts are at their (MMC) when they are at their largest size.

As the tapped hole in Figure 13 departs from maximum material condition (or gets larger), a bonus tolerance is added to the 0.25 mm equal to the amount of the departure. The bonus tolerance cannot exceed the feature tolerance.

The next control in the feature control frame is "P" and "14," meaning *projected tolerance zone* of 14 mm. The centerline of the tapped hole is projected the distance of 14 mm above the part to create an imaginary cylinder which must contain the centerline of the feature.

The position of the 6 mm tapped hole is with respect to datums A, B, and C. Datum A is shown as a surface, datums B and C are not shown, but are typically two edges of the part.

Dimensioning, Gaging, and Measuring

4x M6x1-6H

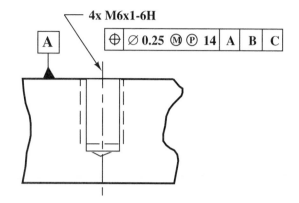

This on the drawing

Means this

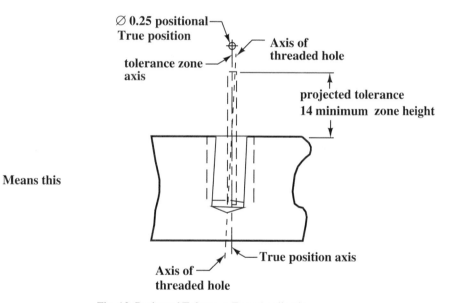

Fig. 13. Projected Tolerance Zone Application.

If this feature control frame read like a story, it would look like this:

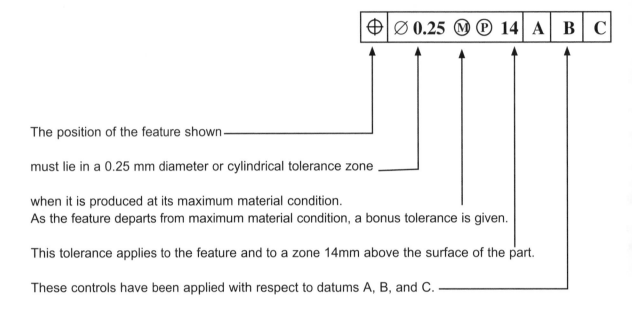

The position of the feature shown

must lie in a 0.25 mm diameter or cylindrical tolerance zone

when it is produced at its maximum material condition.
As the feature departs from maximum material condition, a bonus tolerance is given.

This tolerance applies to the feature and to a zone 14mm above the surface of the part.

These controls have been applied with respect to datums A, B, and C.

Using the Datums

When a machinist clamps a rectangular block in a machine vise on two parallels (supports), the primary datum is satisfied because the base of the block is controlled by three (or more) points. The solid jaw of the vise prevents rotational movement, so the secondary datum has been accomplished. A positive stop used on the vise jaw to act as a "bump stop" acts as the tertiary datum. If multiple parts are required, every part will nest the same way. *The machinist uses the datums shown on the engineering drawing to set up the machine tool. These are the same datums the inspector will use to check the part.*

Unit B: Allowances and Tolerances for Fits pages 628–674

Shop Recommended page 630

Allowance for Forced Fits.—The allowance per inch of diameter usually ranges from 0.001 inch to 0.0025 inch (0.0254-0.0635 mm), 0.0015 inch (0.0381 mm) being a fair average. Ordinarily the allowance per inch decreases as the diameter increases; thus the total allowance for a diameter of 2 inches (50.8 mm) might be 0.004 inch (0.102 mm), whereas for a diameter of 8 inches (203.2 mm) the total allowance might not be over 0.009 or 0.010 inch (0.23 or 0.25 mm). The parts to be assembled by forced fits are usually made cylindrical, although sometimes they are slightly tapered. Advantages of the taper form are: the possibility of abrasion of the fitted surfaces is reduced; less pressure is required in assembling; and parts are more readily separated when renewal is required. On the other hand, the taper fit is less reliable, because if it loosens, the entire fit is free with but little axial movement. Some lubricant, such as white lead and lard oil mixed to the consistency of paint, should be applied to the pin and bore before assembling, to reduce the tendency toward abrasion.

A force fit involving a shaft and hole is a condition where the shaft is larger than the hole. Force fits are also known as interference fits and press fits. The amount of interference between mating parts is critical. The assembly of parts having an interference fit is accomplished with:

- An arbor press
- A soft punch and hammer
- A vise or other screw device
- Liquid nitrogen to shrink the male component
- Heat to increase the size of the female component
- Combinations of the methods described above

The process of pressing bushings into holes is very common in the machine shop. Hardened bushings are used to:

- Provide a hardened hole location for tooling components.
- Guide a cutting tool such as a drill in a drill jig.
- Provide a reliable "zero" location.

Navigation Hint: Information on interference fits and clearance fits is not found in the MH index under *interference* or *clearance*. When this occurs, find another key word. A search of the word "*fits*" reveals the desired information:

- fits/clearance fits/pages 653–654, 657–658
- fits/interference/pages 635, 652

The Carr Lane Manufacturing catalog is an excellent source for technical information on the installation of bushings. These size recommendations can be applied to other press-fit applications. Note the sizes and tolerances of bushings and hole diameters (see Figure 4.8). The "Actual" column shows the guaranteed size of the bushing as furnished by Carr Lane. The "Recommended Hole Size" is the machined size of the hole. The total tolerance for most of the hole diameters is .0003 (pronounced *three tenths*). Great care must be taken when machining holes to this degree of accuracy.

4

SECTION 4

Navigation Hint: For more information on degrees of accuracy, see Section 6,

— Unit J, Machine Tool Accuracy, in this guide, pages 155

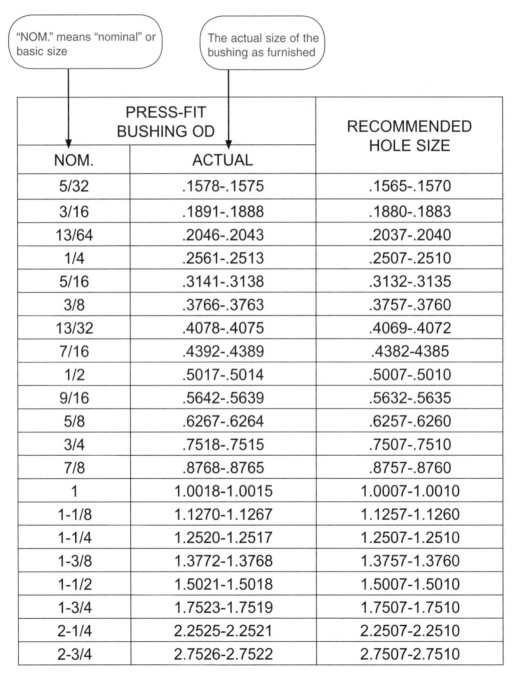

NOM.	PRESS-FIT BUSHING OD	RECOMMENDED HOLE SIZE
	ACTUAL	
5/32	.1578-.1575	.1565-.1570
3/16	.1891-.1888	.1880-.1883
13/64	.2046-.2043	.2037-.2040
1/4	.2561-.2513	.2507-.2510
5/16	.3141-.3138	.3132-.3135
3/8	.3766-.3763	.3757-.3760
13/32	.4078-.4075	.4069-.4072
7/16	.4392-.4389	.4382-4385
1/2	.5017-.5014	.5007-.5010
9/16	.5642-.5639	.5632-.5635
5/8	.6267-.6264	.6257-.6260
3/4	.7518-.7515	.7507-.7510
7/8	.8768-.8765	.8757-.8760
1	1.0018-1.0015	1.0007-1.0010
1-1/8	1.1270-1.1267	1.1257-1.1260
1-1/4	1.2520-1.2517	1.2507-1.2510
1-3/8	1.3772-1.3768	1.3757-1.3760
1-1/2	1.5021-1.5018	1.5007-1.5010
1-3/4	1.7523-1.7519	1.7507-1.7510
2-1/4	2.2525-2.2521	2.2507-2.2510
2-3/4	2.7526-2.7522	2.7507-2.7510

Figure 4.8 "*Compliments of Carr Lane Mfg. Co.*"

Dimensioning, Gaging, and Measuring

Unit C: Measuring Instruments and Inspection Methods
pages 675–732

Index navigation path and key words:

According to the table provided by Carr-Lane, a 1/2″ O. D. (outside diameter) bushing is furnished at .5017 to .5014. The recommended hole size is from .5007 to .5010. That means that the greatest amount of interference between the bushing and the hole is largest bushing size minus the smallest hole size or:

.5017	bushing
−.5007	hole
.0010	interference

The least amount of interference occurs when the bushing is at its smallest diameter and the hole is at its largest diameter:

.5014	bushing
−.5010	hole
.0004	interference

Both possibilities are an acceptable amount of interference for tooling components of this size.

If the interference is too great, the inside diameter of the bushing will collapse, resulting in problems with the fit of mating part. If the interference is insufficient, the bushing will not be held securely.

Micrometers

A micrometer is a precision measuring tool. The part to be measured is placed in between the anvil and the spindle. The thimble is advanced until there is a light drag between the measurement faces and the part being measured. The ratchet knob slips when the proper amount of measurement pressure has been applied.

Close the measurement faces on a clean business card and gently pull on the card to clean the faces. The micrometer shown in Figure 4.9 is a metric micrometer.

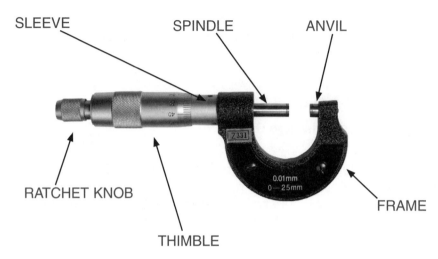

SLEEVE   SPINDLE   ANVIL

RATCHET KNOB

THIMBLE

FRAME

Figure 4.9 Micrometer

The descrimination of a precision measuring instrument is the degree to which it divides measuring units. An inch or "English" micrometer divides an inch into one thousand or ten thousand equal parts, depending on the type of micrometer. The bold numbers on the sleeve indicate divisions of .100 (one hundred thousandths). The micrometer is read by observing the part of the sleeve that is not covered up by the thimble. Learning to read an inch micrometer is made easier by using dollars, quarters, and pennies as indicators. Every bold number on the sleeve may be considered a "dollar". There are four equal spaces in between each bold number (dollar) which can be considered "quarters". The number on the thimble, or rotating part, is pennies. See Figure 4.10.

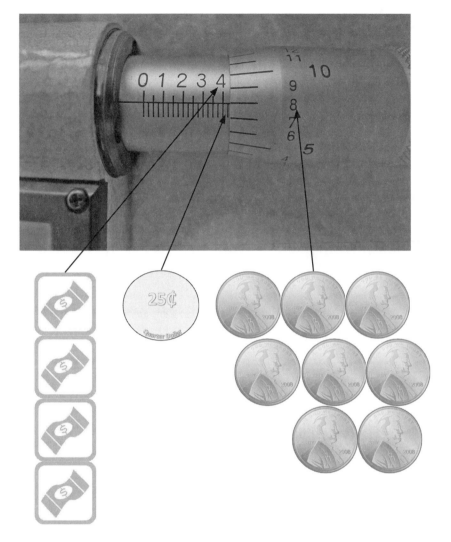

$4.00 + .25 + .008 = $4.33
Move the decimal point one place to the left for a reading of .433.

Figure 4.10 An inch micrometer

To read the micrometer directly, use the following method: The number "4" is exposed, so the reading is .400 (four hundred thousandths) + one twenty-five thousandths line (.025) + eight thousandths shown on the thimble (.008) for a reading of four hundred thirty-three thousandths.

$$
\begin{array}{r}
.400 \\
.025 \\
\underline{.008} \\
.433
\end{array}
$$

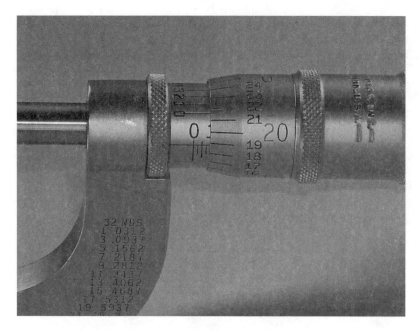

Figure 4.11 Reading a micrometer

The reading on the micrometer in Figure 4.11 is:

Three twenty-five thousandths lines (.075) +
Nineteen thousands on the thimble (.019) =
.094

Vernier Scales

Some micrometers have a vernier scale on the frame so that measurements within 0.0001 (one tenth) can be taken (see Figure 4.12). First determine the reading in thousandths, and then find the line on the vernier scale that lines up with the line on the thimble. Add this number to the end of the thousandths place in the fourth position. If the micrometer does not have a vernier scale, the reading is taken at the closest line.

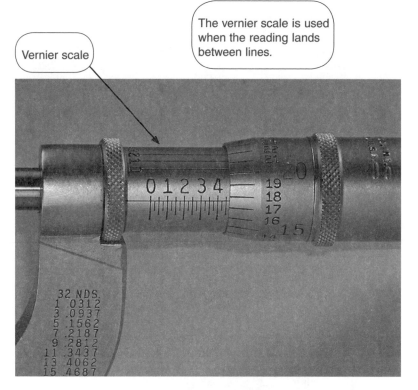

Figure 4.12 Vernier scale

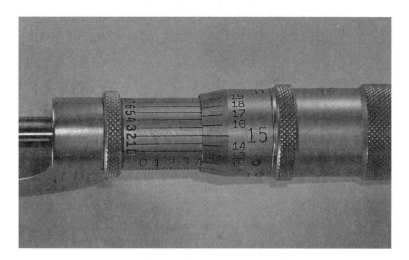

Figure 4.13 Reading a vernier scale

There are ten lines on the vernier scale which correspond to nine lines on the thimble, so only one position can be aligned. In Figure 4.13, the "2" lines up with the thimble lines, so .0002 is added to the thousandths reading. When reading the "tenth" scale, the numbers on the thimble are disregarded; only the lines on the sleeve are used for the reading.

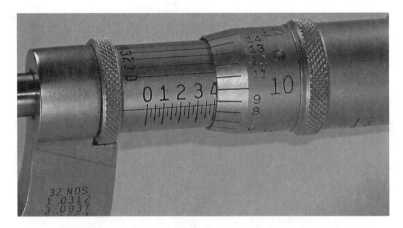

Figure 4.14a Four major divisions plus eight divisions on the thimble (almost 9)

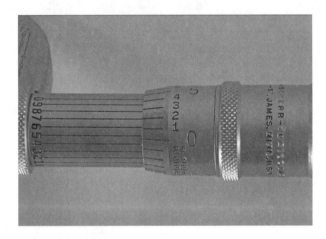

Figure 4.14b. The "7" on the vernier scale lines up

In Figure 4.14, four major divisions are exposed showing a reading of .400 plus eight divisions on the circumference of the thimble for an additional .008. If this micrometer did not have a vernier scale, the reading would be .409 because the nine is the closest line. The reading is not quite .409, and because this is a "tenth mic," it can be read to the fourth place. Add the reading from the vernier scale to .408 for .4087.

Reading a vernier caliper

The vernier caliper is used in situations where a steel rule is not accurate enough. Read the instrument at the zero line by observing the increments to the left of the line (see Figure 4.15). This tool reads in metric on top scale and in inches on the bottom scale. There are ten equal divisions between each inch mark or .100 or $\frac{1}{10}$. The .100 lines are further divided into four equal spaces or .025 each. The rest of the reading is observed on the vernier scale. Only one line on the vernier scale will line up with the line above it because the lines are spaced differently.

In Figure 4.15, determine the reading by observing the scale to left of the zero line. The 1 is exposed for reading of: 1.000

Two one hundred thousandths lines are exposed for a reading of: .200

Add to this *two* twenty-five thousandth lines for a reading of: .050

The rest of the reading is taken from the vernier scale. Of the numbers on the lower scale, the four is coincident, or at the same spot, as the line above it. Add .004   .004
                                           The total reading is: 1.254

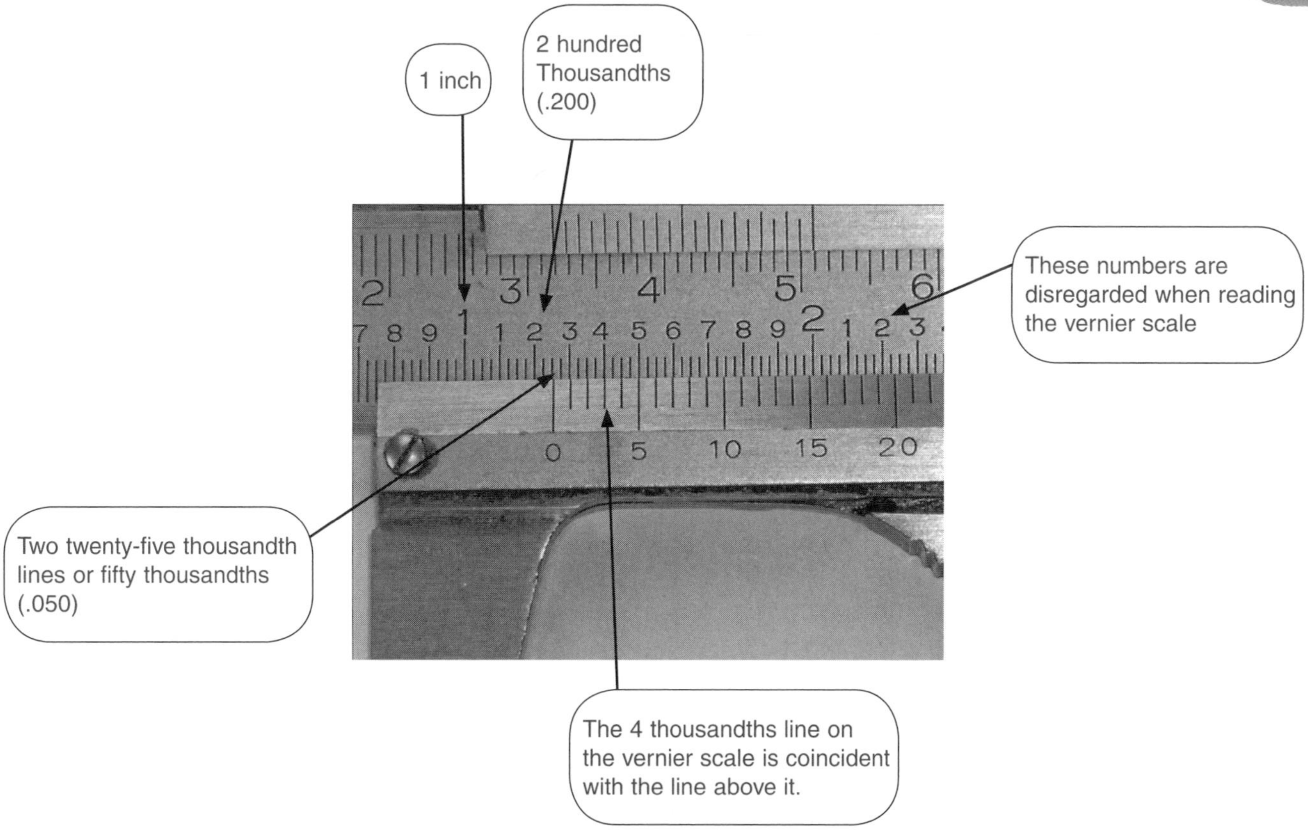

Figure 4.15 Reading a vernier caliper

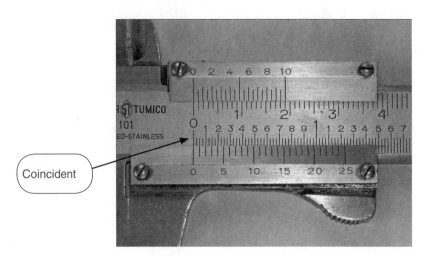

Figure 4.C.9 16 Coincident zero lines.

Before using a vernier caliper, close it and make sure the zero lines are coincident (see Figure 4.16).

Unit D: Surface Texture

Designers specify the smoothness of a surface based on the type of service, function, and operating conditions of the part or assembly. A symbol is used to convey the "roughness" requirement of the surface as seen in the table on page 742. The following symbols are used to convey finish requirements.

Dimensioning, Gaging, and Measuring

Shop Recommended page 742

Surface Texture Symbols and Construction

Symbol	Meaning
√ Fig. 7a.	Basic Surface Texture Symbol. Surface may be produced by any method except when the bar or circle (Fig. 7b or 7d) is specified.
⩔ Fig. 7b.	Material Removal By Machining Is Required. The horizontal bar indicates that material removal by machining is required to produce the surface and that material must be provided for that purpose.
3.5 ⩔ Fig. 7c.	Material Removal Allowance. The number indicates the amount of stock to be removed by machining in millimeters (or inches). Tolerances may be added to the basic value shown or in general note.
⩗ Fig. 7d.	Material Removal Prohibited. The circle in the vee indicates that the surface must be produced by processes such as casting, forging, hot finishing, cold finishing, die casting, powder metallurgy or injection molding without subsequent removal of material.
√⎯ Fig. 7e.	Surface Texture Symbol. To be used when any surface characteristics are specified above the horizontal line or the right of the symbol. Surface may be produced by any method except when the bar or circle (Fig. 7b and 7d) is specified.

Index navigation path and key words:

— surface texture/applying symbols/page 742

Shop Recommended page 743, Figure 8

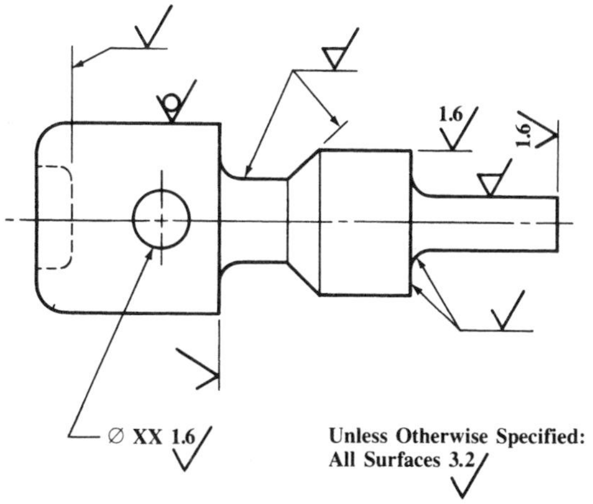

Figure 8. Application of surface texture symbols

Processes and Surface Finishes

Different production methods result in different surface finishes. The skilled machinist selects the machine tool that is the most efficient *and* produces the required surface finish.

Reasons for rough finishes:

- The surface is a glue joint.
- The surface is unimportant.
- An as-cast finish is desirable.

Reasons for smooth finishes:

- A high degree of accuracy is required. Close tolerances cannot be held with rough finishes.
- The quality of the piece part depends on the surface finish of the tool.
- Appearance.
- Flow of a solid or a liquid is influenced by the smoothness and direction of polish.

4

Shop Recommended page 739, Table 1

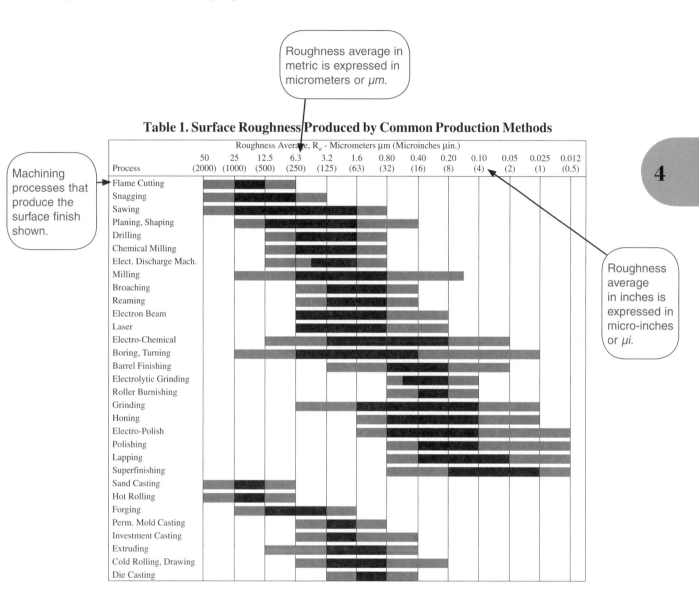

Table 1. Surface Roughness Produced by Common Production Methods

Roughness average in metric is expressed in micrometers or *μm*.

Machining processes that produce the surface finish shown.

Roughness average in inches is expressed in micro-inches or *μi*.

4

Navigation Hint: Surface texture and machine tool accuracy are directly related. For information on machine tool accuracy, see Section 6; Machining Operations; in *Machinery's Handbook Made Easy*, page 155;

SECTION 4

Shop Recommended page 745, Table 5

In addition to roughness average, other controls may be added to the finish symbol. One example is Lay Symbols. This refers to the direction of the finish, similar to a wood grain. To satisfy a Lay Symbol requirement, it may be necessary to polish the surface in the required direction.

Table 5. Lay Symbols

Lay Symbol	Meaning	Example Showing Direction of Tool Marks
=	Lay approximately parallel to the line representing the surface to which the symbol is applied.	
⊥	Lay approximately perpendicular to the line representing the surface to which the symbol is applied.	
X	Lay angular in both directions to line representing the surface to which the symbol is applied.	
M	Lay multidirectional	
C	Lay approximately circular relative to the center of the surface to which the symbol is applied.	
R	Lay approximately radial relative to the center of the surface to which the symbol is applied.	
P	Lay particulate, non-directional, or protuberant	

Shop Recommended page 746 Table 6,

Table 6. Application of Surface Texture Values to Symbol *ANSI B46.1-1978*

1.6 ∨	Roughness average rating is placed at the left of the long leg. The specification of only one rating shall indicate the maximum value and any lesser value shall be acceptable. Specify in micrometers (microinch).
1.6 **0.8** ∨	The specification of maximum and minimum roughness average values indicates permissible range of roughness. Specify in micrometers (microinch).
0.005-5 **0.8** ∨	Maximum waviness height rating is the first rating place above the horizontal extension. Any lesser rating shall be acceptable. Specify in millimeters (inch). Maximum waviness spacing rating is the second rating placed above the horizontal extension and to the right of the waviness height rating. Any lesser rating shall be acceptable. Specify in millimeters (inch).
1.6 **3.5** ∨	Material removal by machining is required to produce the surface. The basic amount of stock provided for material removal is specified at the left of the short leg of the symbol. Specify in millimeters (inch).
1.6 ∨	Removal of material is prohibited.
0.8 ∨⊥	Lay designation is indicated by the lay symbol placed at the right of the long leg.
0.8 **2.5** ∨	Roughness sampling length or cutoff rating is placed below the horizontal extension. When no value is shown, 0.80 mm (0.030 inch) applies. Specify in millimeters (inch).
0.8 ⊥ **0.5** ∨	Where required maximum roughness spacing shall be placed at the right of the lay symbol. Any lesser rating shall be acceptable. Specify in millimeters (inch).

The values shown in this table are metric.

SECTION 4

ASSIGNMENT

List the key terms and give a definition of each.

Meter	Hidden Line	Tolerance
Millimeter	Center Line	Discrimination
Inch	Datum	Maximum Material Condition
Force Fit	Basic Dimensions	
Visible line	Interference Fit	

4

APPLY IT!

Review questions for Dimensioning, Gaging, and Measuring.

1. How many threads per inch are on a 0-1″ micrometer?

2. What is the reading (in inches) on the vernier calipers shown below?

a.

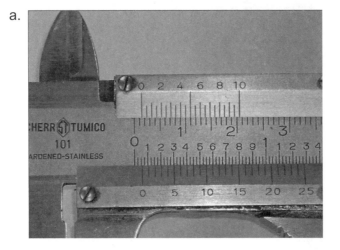

b.

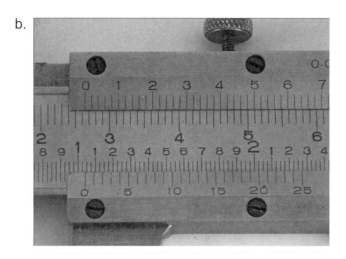

c.

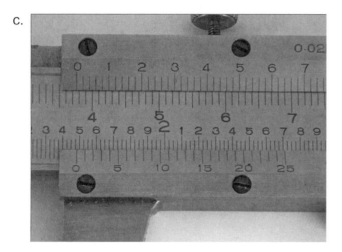

3. What are the readings on the micrometers shown below?

a.

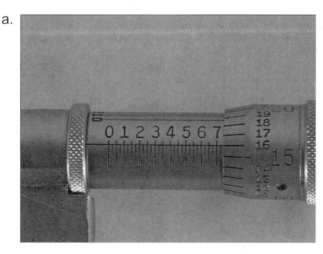

b.

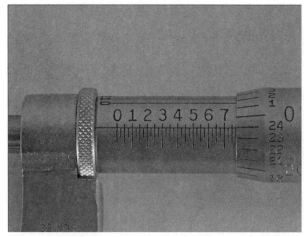

c.

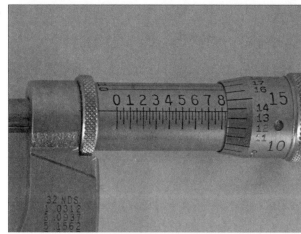

d. Use the vernier scale to give the following reading.

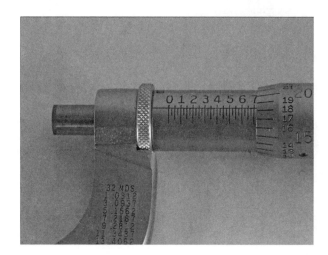

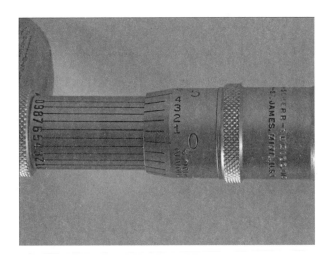

4

4. A standard 0-1 inch micrometer divides an inch into how many equal pieces?

5. What types of machine tools are capable of producing the following surface finishes (roughness average in µin.)?

   a. 125

   b. 16

   c. 250

   d. 8

   e. 63

6. According to the Carr Lane technical information table, what is the maximum amount press fit allowable for a:

   a. 1/2″ O.D. bushing

   b. 5/16″ O.D. bushing

   c. 7/8″ O.D. bushing

7. According to the Carr Lane technical information table, what is the minimum amount press fit allowable for a:

   a. 3/8″ O.D. bushing

   b. 7/16″ O.D. bushing

   c. 1 1/2″ O.D. bushing

8. Make a sketch of a part provided by your instructor. Use proper drafting techniques.

9. Obtain an engineering drawing that has GD&T controls. In small groups, interpret the feature control frames.

Section 5

TOOLING and TOOLMAKING

Use this section with pages 753 — 1003
in Machinery's Handbook 29th Edition

SECTION 5

Key Terms

Positive Rake	Tapered Shank
Negative Rake	Flute
Nose Radius	Effective Thread
High Speed Steel	Pitch
Reaming Allowance	Major Diameter
Helical Flute	Minor Diameter
Shank	Blind Hole
Point Angle	Cratering
Margin	

88

Learning Objectives

After studying this unit you should be able to:

- List six different operations that high speed steel cutters can produce.
- Describe the difference between positive rake cutting tools and negative rake cutting tools.
- Define the characteristics of the ten positions used to identify carbide inserts.
- Define the tool holders used to machine a variety of features.
- List the advantages and disadvantages of three different carbide insert shapes.
- Select the correct clearance drill for a specific screw.
- List the four systems of drill sizes.
- Use the tables from the Machinery's Handbook to select a drill.
- Select the proper drill for a specific tap.
- Calculate the correct drilled depth for a tapped hole.
- State the most common reasons for tap breakage.
- State the parts of a screw thread designation.
- Calculate the sizes of the drills used to prepare for the reaming operation.
- List the three systems used to identify reamer sizes.

5

Introduction

Tools for turning, milling, slitting, contour machining, etc., are made in a variety of shapes and are held in special holders to present the cutting edges to the surface of the workpiece. Different tool shapes and tool contours have an effect on machining efficiency. Review the terms beginning on page 757 of *Machinery's Handbook 29*. Tool Wear and Sharpening (Unit L) in this book copies selected excerpts from MH 29 followed by a bulleted list of simplified explanations.

Unit A: Cutting Tools pages 757–771

This section of *Machinery's Handbook* focuses on basic metal cutting tools. For a tool to cut metal, the tool must be harder than the workpiece. Thousands of variations of cutting tools are available for today's metal cutting industry. There are two major categories:

- High Speed Steel
- Sintered and Cemented Carbide

High Speed Steel

High speed steel cutting tools are considered a *super alloy* because this material has a large percentage of alloying elements. High speed steels are divided into two major groups: *tungsten type* and *molybdenum types*. The principal properties of these steels are the ability to withstand high temperatures and abrasion and retain their hardness deeply into the cutting tool. Most high speed tools are used in turning machines such as lathes. The tool may assume many different shapes based on the machining operation. Unlike other harder cutting tools, high speed steel cutting tools can be easily shaped on a pedestal grinder with an aluminum-oxide grinding wheel. This process is known as "off-hand grinding". High speed steel cutting tools are not usually used for production or for machining tough, high-alloy steels.

 Navigation Hint: For more information on alloys, see Section 3, Properties, Treatment, and Testing of Materials MHB29 pages 369–607

Turning operations that use high speed steel cutting tools are:

- Straight turning — Reducing the diameter of a shaft
- Facing — Cutting a flat face on the end of a shaft
- Threading — Cutting a series of grooves in a shaft to produce a screw thread
- Grooving — Plunging into the outside diameter of a shaft to produce an undercut
- Boring — Enlarging a previously drilled hole by taking a series of cuts on the inside diameter
- Parting off — Cutting off a portion of a shaft with a blade-shaped tool by plunging

Figure 5.1 illustrates several of these operations.

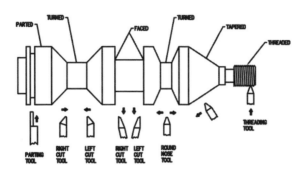

Figure 5.1 Examples of operations done by high speed steel cutting tools

Refer to MH 29, pages 757–771 for details on tool geometry and definitions. Different angles shown for the cutting tools give the cutter strength, clearance, chip flow, chip formation, surface finish, tool life, and machinability. It should be noted that all angles of the tool slope away from the cutting edge, but at various angles. Of the terms describing tool angles and geometry, the ones that affect the machining process the most are *positive rake, negative rake,* and *nose radius* (see Figure 5.2).

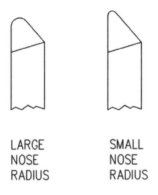

LARGE   SMALL
NOSE   NOSE
RADIUS   RADIUS

Figure 5.2

Positive rake has these advantages:

- Less cutting force required
- Lower cutting temperature
- Longer tool life

Negative Rake has these advantages:

- Absorbs shock for interrupted cuts
- Top and bottom of the inserts can used for double the tool life

The *nose radius* of the tool refers to the rounded point of the tool that makes contact with the workpiece. Use the largest nose radius possible. Some benefits of a large nose radius are:

- Better surface finishes
- Stronger than small nose radius
- Makes a thinner chip for better heat dissipation

However, if the nose radius is excessive, it can result in vibration or "chatter" because of increased tool to workpiece contact. A smaller nose radius has the opposite effect of the benefits listed above.

These tools are sometimes referred to as "single point cutting tools" because the cutting portion of the tool contacts the workpiece at one point.

 Index navigation paths and key words:

— cutting tools/angles/pages 757–772, 782–784
— cutting tools/high speed steel/page 1009

Unit B: Cemented Carbides pages 785–795

Cemented carbides are also known as *sintered carbides* or just *carbide*. These cutters have superior wear resistance and produce good surface finishes. *Cemented carbides* are much harder and more brittle than high speed steel tools. Spindle speeds of three-to-five times greater than high speed steel tools are typical. Carbide cutting tools come in the form of an insert that is held in a tool holder. Solid carbide milling cutters are also available. Inserts that have multiple cutting edges are known as *indexable*. When a cutting edge becomes dull, the insert is "indexed" to expose a new cutting edge. Inserts are sometimes known as "throw away" inserts because they are not normally sharpened; they are recycled. The American International Standards Institute (ANSI) has standardized the identification system that describes the inserts and the special tool holders required.

This is the 10-digit identification system used to choose an insert. Every position defines a characteristic of the insert.

1	2	3	4	5	6	7	8	9	10
T	N	M	G	5	4	3			A

Tooling and Toolmaking

Companies that sell carbide inserts use this system. Here is a summary of how the system works.

1. Shape of the insert
2. Relief angle: The angular clearance below the cutting edge
3. Tolerance: When the insert is indexed, tolerance is the ability of the newly indexed cutting edge to be at the same location as the previous edge. This ability is important because it is common for the machine tool to have the end of the insert referenced to the workpiece.
4. This field is for insert shape options and how the insert is mounted in the holder. Options of insert types are with or without hole, with recessed fastener hole, with chip groove, and other configurations.
5. Number of eighths of an inch in the inscribed circle of the insert (see Figure 5.3).

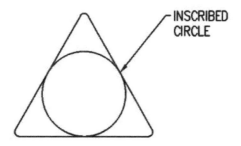

INSCRIBED CIRCLE

Figure 5.3 Inscribed circle

6. Number of sixteenths of an inch in the insert's thickness.
7. Cutting point configuration. Numbers are for a radius; letters are for a chamfer or angle. Zero is sharp; 1 is 1/64 inch radius, 2 is 2/64 (1/32) radius and so on. Letter A is for a square insert with a 45-degree angle; letter D is for a square insert with a 30-degree angle.
8. Special cutting point definition. This position is used when the field 7 is represented by a letter. It is the number of 64ths in the size of the facet. May or may not be included depending on the type of insert.
9. Right or left hand. May or may not be included depending on the type of insert.
10. Other conditions. This position is to identify any special surface treatments for the insert such as honed or polished. Honed or polished surfaces improve chip flow.

Identify the insert described below:

1	2	3	4	5	6	7	8	9	10
T	N	M	G	5	4	3			A

The insert identified above is:

T	triangular
N	0 degree relief angle
M	tolerance of .002 to .004 for inscribed circle
G	chip grooves with a hole
5	number of eighths of an inch in the inscribed circle (5/8)
4	number of sixteenths of an inch in the thickness (4/16 or 1/4)
3	number of sixty-fourths of an inch in the tip radius (3/64)
	— no special cutting point
	— not left or right hand
A	special conditions; honed

Check these pages for more information:

— Page 765 of MH29 has more details of each category.
— Also, the suppliers of carbide inserts have a portion of their catalog dedicated to the insert identification system.

Three Popular Insert Shapes

Round inserts have the most strength and greatest number of cutting edges because rotating the insert slightly can expose a new cutting edge. Round insert are good for straight turns on the outside diameter of a shaft. This shape, however, cannot be used to machine corners or small undercuts.

Triangular inserts are very versatile. These inserts can machine areas of the workpiece that other shapes can't access, such as turning an outside diameter up to a square corner.

Square inserts are stronger than triangular inserts and have more cutting edges than triangular inserts, but they cannot be used to produce a square shoulder on a round shaft. An insert that has corner angles of less than 90 degrees is required for that operation.

> *Analyze, Evaluate, & Implement*

Use a commercial cutting tool catalog to select a carbide cutter for a specific machining operation.

Tool Holders

Indexable inserts require special tool holders. The insert fits into a pocket on the holder and may be held by number of different clamping methods. The holder can present the cutting tool in one of three different *rake angles*: positive, negative, or neutral.

Index navigation paths and key words:

Shop Recommended pages 770–772, Table 3b

Table 3b. Indexable Insert Holder Application Guide

Tool	Tool Holder Style	Insert Shape	N-Negative P-Positive Rake	Turn	Face	Turn and Face	Turn and Backface	Trace	Groove	Chamfer	Bore	Plane
	A	T	N	•	•						•	
			P	•	•						•	
	A	T	N	•	•			•				
			P	•	•			•				
	A	R	N	•	•	•						•
	A	R	N	•	•	•		•				•
	B	T	N	•	•						•	
			P	•	•						•	
	B	T	N	•	•			•			•	
			P	•	•			•			•	
	B	S	N	•	•						•	
			P	•	•						•	
	B	C	N	•	•	•					•	•
	C	T	N	•	•				•	•		
			P	•	•				•	•		

> The Application section of Table 3b shows the type of machining operation the toolholder and insert is capable of. These applications refer to a turning machine.
>
> **Turn** is reducing the outside diameter of the workpiece.
>
> **Face** is cutting a flat surface on the end of the workpiece.
>
> **Backface** is cutting a flat surface on the bottom of the inside of a hole.

5

Reamers

Reaming is a secondary operation that slightly enlarges a previously drilled hole. Reamed holes are very accurate and have a superior surface finish.

Reaming should be done with cutting oil unless the material is cast iron, bronze, or brass. These materials are normally machined dry. Speeds and feeds for reaming depend on material, required accuracy, and other factors, but as a general rule reamers are operated at half the speed of drilling and twice the feed. The most popular types of reamers are:

- Expansion hand reamers
- Hand reamers
- Chucking reamers

Expansion hand reamers have a slightly adjustable diameter. These reamers are used to increase the diameter of a hole by about .001 to .003. The center of the reamer is hollow. It has a tapered section that causes the outside diameter of the reamer to increase by twisting a threaded screw.

Hand reamers have either straight flutes or spiral flutes similar to a twist drill. The spiral fluted reamers are useful for bridging a small interrupted section of a hole such as a keyway. The square end is driven by a tap wrench or similar tool.

Chucking reamers are driven by machine tools such as drill presses, lathes, and milling machines.

The *reaming allowance* is amount of material left in the hole after drilling. Keep in mind that a properly sharpened drill produces a hole that is about .002 to .008 oversize. For high accuracy, a common practice is to drill the hole undersize, then drill the hole again using a larger drill to obtain the proper amount of reaming allowance.

Shop Recommended page 897

Oversize Diameters in Drilling

Drill Dia., Inch	Amount Oversize, Inch			Drill Dia., Inch	Amount Oversize, Inch		
	Average Max.	Mean	Average Min.		Average Max.	Mean	Average Min.
1/16	0.002	0.0015	0.001	1/2	0.008	0.005	0.003
1/8	0.0045	0.003	0.001	3/4	0.008	0.005	0.003
1/4	0.0065	0.004	0.0025	1	0.009	0.007	0.004

Use the following chart to determine the drill size used prior to reaming the hole.

Size of Reamer	Reaming Allowance
1/32 to 1/8	.003 to .005
1/8 to 1/4	.004 to .008
1/4 to 3/8	.006 to .010
3/8 to 1/2	.010 to .015
1/2 to 3/4	.015 to .03
3/4 to 1.00	.03

Analyze, Evaluate, & Implement

Make a tool list to machine ream a:

- *.3755 dia. hole*
- *.120 dia. hole*
- *.250 dia. hole*

Tooling and Toolmaking

Shop Recommended Reamers page 858

Straight Shank Chucking Reamers—Straight Flutes, Wire Gage Sizes
ANSI B94.2-1983 (R1988)

Chucking reamers are machine driven as opposed to hand operated.

Note that the shank diameter is slightly less than the reamer diameter so that the reamer can pass through the hole.

The first column shows the wire gage size followed by the diameter in thousandths of an inch.

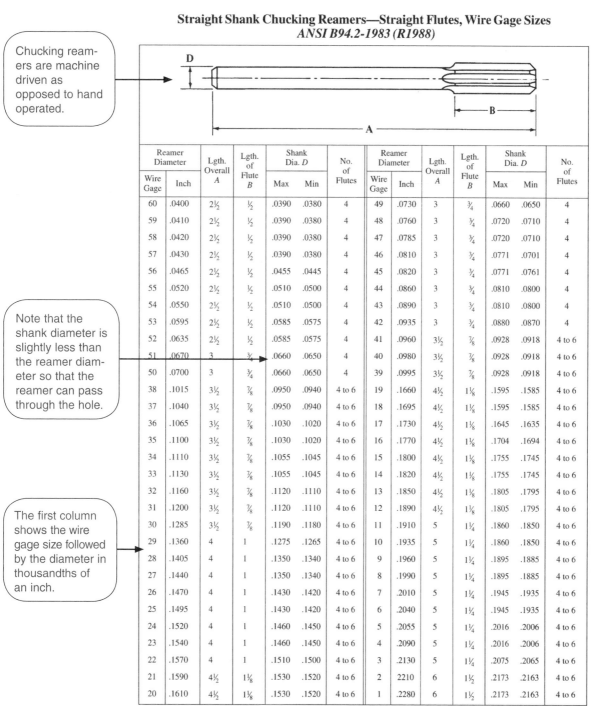

Reamer Diameter		Lgth. Overall A	Lgth. of Flute B	Shank Dia. D		No. of Flutes	Reamer Diameter		Lgth. Overall A	Lgth. of Flute B	Shank Dia. D		No. of Flutes
Wire Gage	Inch			Max	Min		Wire Gage	Inch			Max	Min	
60	.0400	2½	½	.0390	.0380	4	49	.0730	3	¾	.0660	.0650	4
59	.0410	2½	½	.0390	.0380	4	48	.0760	3	¾	.0720	.0710	4
58	.0420	2½	½	.0390	.0380	4	47	.0785	3	¾	.0720	.0710	4
57	.0430	2½	½	.0390	.0380	4	46	.0810	3	¾	.0771	.0701	4
56	.0465	2½	½	.0455	.0445	4	45	.0820	3	¾	.0771	.0761	4
55	.0520	2½	½	.0510	.0500	4	44	.0860	3	¾	.0810	.0800	4
54	.0550	2½	½	.0510	.0500	4	43	.0890	3	¾	.0810	.0800	4
53	.0595	2½	½	.0585	.0575	4	42	.0935	3	¾	.0880	.0870	4
52	.0635	2½	½	.0585	.0575	4	41	.0960	3½	⅞	.0928	.0918	4 to 6
51	.0670	3	¾	.0660	.0650	4	40	.0980	3½	⅞	.0928	.0918	4 to 6
50	.0700	3	¾	.0660	.0650	4	39	.0995	3½	⅞	.0928	.0918	4 to 6
38	.1015	3½	⅞	.0950	.0940	4 to 6	19	.1660	4½	1⅛	.1595	.1585	4 to 6
37	.1040	3½	⅞	.0950	.0940	4 to 6	18	.1695	4½	1⅛	.1595	.1585	4 to 6
36	.1065	3½	⅞	.1030	.1020	4 to 6	17	.1730	4½	1⅛	.1645	.1635	4 to 6
35	.1100	3½	⅞	.1030	.1020	4 to 6	16	.1770	4½	1⅛	.1704	.1694	4 to 6
34	.1110	3½	⅞	.1055	.1045	4 to 6	15	.1800	4½	1⅛	.1755	.1745	4 to 6
33	.1130	3½	⅞	.1055	.1045	4 to 6	14	.1820	4½	1⅛	.1755	.1745	4 to 6
32	.1160	3½	⅞	.1120	.1110	4 to 6	13	.1850	4½	1⅛	.1805	.1795	4 to 6
31	.1200	3½	⅞	.1120	.1110	4 to 6	12	.1890	4½	1⅛	.1805	.1795	4 to 6
30	.1285	3½	⅞	.1190	.1180	4 to 6	11	.1910	5	1¼	.1860	.1850	4 to 6
29	.1360	4	1	.1275	.1265	4 to 6	10	.1935	5	1¼	.1860	.1850	4 to 6
28	.1405	4	1	.1350	.1340	4 to 6	9	.1960	5	1¼	.1895	.1885	4 to 6
27	.1440	4	1	.1350	.1340	4 to 6	8	.1990	5	1¼	.1895	.1885	4 to 6
26	.1470	4	1	.1430	.1420	4 to 6	7	.2010	5	1¼	.1945	.1935	4 to 6
25	.1495	4	1	.1430	.1420	4 to 6	6	.2040	5	1¼	.1945	.1935	4 to 6
24	.1520	4	1	.1460	.1450	4 to 6	5	.2055	5	1¼	.2016	.2006	4 to 6
23	.1540	4	1	.1460	.1450	4 to 6	4	.2090	5	1¼	.2016	.2006	4 to 6
22	.1570	4	1	.1510	.1500	4 to 6	3	.2130	5	1¼	.2075	.2065	4 to 6
21	.1590	4½	1⅛	.1530	.1520	4 to 6	2	2210	6	1½	.2173	.2163	4 to 6
20	.1610	4½	1⅛	.1530	.1520	4 to 6	1	.2280	6	1½	.2173	.2163	4 to 6

All dimensions in inches. Material is high-speed steel.
Tolerances: On diameter of reamer, plus .0001 to plus .0004 inch. On overall length *A*, plus or minus ¹⁄₁₆ inch. On length of flute *B*, plus or minus ¹⁄₁₆ inch.

5

97

SECTION 5

Index navigation paths and key words:

— reamers/chucking/pages 846, 849, 853, 858–859

Shop Recommended Reamers page 859

Straight Shank Chucking Reamers—Straight Flutes, Letter Sizes
ANSI B94.2-1983 (R1988)

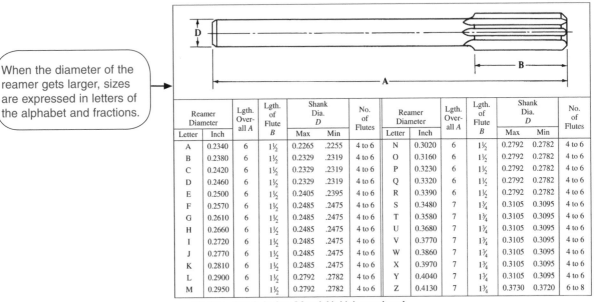

> When the diameter of the reamer gets larger, sizes are expressed in letters of the alphabet and fractions.

Reamer Diameter		Lgth. Over-all A	Lgth. of Flute B	Shank Dia. D		No. of Flutes	Reamer Diameter		Lgth. Over-all A	Lgth. of Flute B	Shank Dia. D		No. of Flutes
Letter	Inch			Max	Min		Letter	Inch			Max	Min	
A	0.2340	6	1½	0.2265	.2255	4 to 6	N	0.3020	6	1½	0.2792	0.2782	4 to 6
B	0.2380	6	1½	0.2329	.2319	4 to 6	O	0.3160	6	1½	0.2792	0.2782	4 to 6
C	0.2420	6	1½	0.2329	.2319	4 to 6	P	0.3230	6	1½	0.2792	0.2782	4 to 6
D	0.2460	6	1½	0.2329	.2319	4 to 6	Q	0.3320	6	1½	0.2792	0.2782	4 to 6
E	0.2500	6	1½	0.2405	.2395	4 to 6	R	0.3390	6	1½	0.2792	0.2782	4 to 6
F	0.2570	6	1½	0.2485	.2475	4 to 6	S	0.3480	7	1¾	0.3105	0.3095	4 to 6
G	0.2610	6	1½	0.2485	.2475	4 to 6	T	0.3580	7	1¾	0.3105	0.3095	4 to 6
H	0.2660	6	1½	0.2485	.2475	4 to 6	U	0.3680	7	1¾	0.3105	0.3095	4 to 6
I	0.2720	6	1½	0.2485	.2475	4 to 6	V	0.3770	7	1¾	0.3105	0.3095	4 to 6
J	0.2770	6	1½	0.2485	.2475	4 to 6	W	0.3860	7	1¾	0.3105	0.3095	4 to 6
K	0.2810	6	1½	0.2485	.2475	4 to 6	X	0.3970	7	1¾	0.3105	0.3095	4 to 6
L	0.2900	6	1½	0.2792	.2782	4 to 6	Y	0.4040	7	1¾	0.3105	0.3095	6 to 8
M	0.2950	6	1½	0.2792	.2782	4 to 6	Z	0.4130	7	1¾	0.3730	0.3720	6 to 8

All dimensions in inches. Material is high-speed steel.

Tolerances: On diameter of reamer, for sizes A to E, incl., plus .0001 to plus .0004 inch and for sizes F to Z, incl., plus .0001 to plus .0005 inch. On overall length A, plus or minus $\frac{1}{16}$ inch. On length of flute B, plus or minus $\frac{1}{16}$ inch.

Straight Shank Chucking Reamers— Straight Flutes, Decimal Sizes
ANSI B94.2-1983 (R1988)

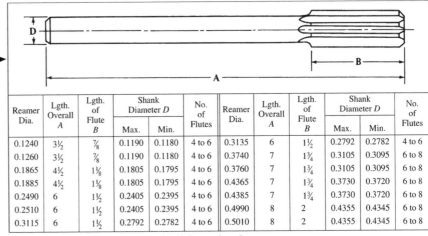

> Note that fractional size reamers are available in fractions of 8ths and 16ths. This is because reamers are typically used to produce holes for dowel pins and other hardware sized the same way.

Reamer Dia.	Lgth. Overall A	Lgth. of Flute B	Shank Diameter D		No. of Flutes	Reamer Dia.	Lgth. Overall A	Lgth. of Flute B	Shank Diameter D		No. of Flutes
			Max.	Min.					Max.	Min.	
0.1240	3½	⅞	0.1190	0.1180	4 to 6	0.3135	6	1½	0.2792	0.2782	4 to 6
0.1260	3½	⅞	0.1190	0.1180	4 to 6	0.3740	7	1¾	0.3105	0.3095	6 to 8
0.1865	4½	1⅛	0.1805	0.1795	4 to 6	0.3760	7	1¾	0.3105	0.3095	6 to 8
0.1885	4½	1⅛	0.1805	0.1795	4 to 6	0.4365	7	1¾	0.3730	0.3720	6 to 8
0.2490	6	1½	0.2405	0.2395	4 to 6	0.4385	7	1¾	0.3730	0.3720	6 to 8
0.2510	6	1½	0.2405	0.2395	4 to 6	0.4990	8	2	0.4355	0.4345	6 to 8
0.3115	6	1½	0.2792	0.2782	4 to 6	0.5010	8	2	0.4355	0.4345	6 to 8

All dimensions in inches. Material is high-speed steel.

Tolerances: On diameter of reamer, for 0.124 to 0.249-inch sizes, plus .0001 to plus .0004 inch and for 0.251 to 0.501-inch sizes, plus .0001 to plus .0005 inch. On overall length A, plus or minus $\frac{1}{16}$ inch. On length of flute B, plus or minus $\frac{1}{16}$ inch.

Counterbores

A counterboring tool is a flat-bottomed cutter with a pilot diameter, driven by the shank end (see Figure 5.4). The pilot diameter is a slip fit into a previously drilled hole. Counterbored holes are usually used to recess a fastener below the surface of the material or to provide a flat machined area (spotface) for a fastener, such as a socket head cap screw. Socket head cap screws are used extensively in industry. Before drilling a hole to accept a counterboring tool, it is important to measure the pilot diameter of the counterbore. Select a drill that is .002 to .005 smaller than the pilot diameter to provide clearance for the tool. Lubricate the pilot of the counterbore to prevent binding.

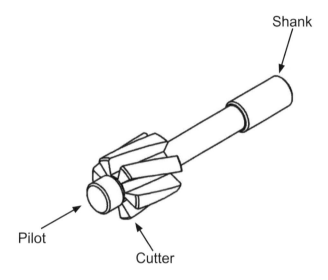

Figure 5.4 Counterboring tool

Figure 5.5 shows a cross-section of a socket head cap screw in a counterbored hole. The hole has been counterbored to a depth that puts the top of the screw .03 beneath the top of the plate.

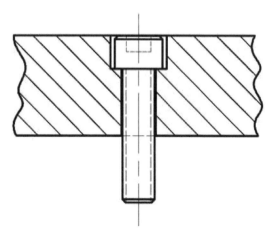

Figure 5.5 Socket head cap screw in assembly

The hole for the fastener must be larger than the diameter of the screw. The difference between the hole size and the diameter of the fastener is the clearance. The amount of clearance depends on the size of the screw. The following clearances are typical.

Fastener	Clearance
Up to 1/4 diameter	.015 or 1/64
1/4 to 1/2	.03 or 1/32
Over 1/2	.06 or 1/16

Analyze, Evaluate, & Implement

Create a table listing the clearance drills for the Socket Head Cap Screws from #8 to 1/2".

Twist Drills

The most common method of producing holes is by drilling. The part of the drill that is held in the machine tool is called the *shank* (see Figure 5.6). Straight shank drills have a uniform cylindrical body for clamping while *taper shank* drills have a conical shaped shank. Large diameter drills (larger than about 1/2") use a taper shank and are driven by the mating tapered holes in the spindle or tool holder. This connection is a locking taper because of the slight angle. The flat *tang* on the small end of the taper is for removing the drill from the holder. A flat, tapered drift is inserted into a slot in the holder and the drill is driven out with a sharp blow from a hammer.

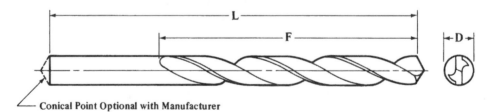

Conical Point Optional with Manufacturer

Figure 5.6 Twist drill

- L = Overall Length
- F = Flute Length
- D = Diameter of Drill

Tooling and Toolmaking

Index navigation paths and key words:

— twist drill/pages 866, 870–896

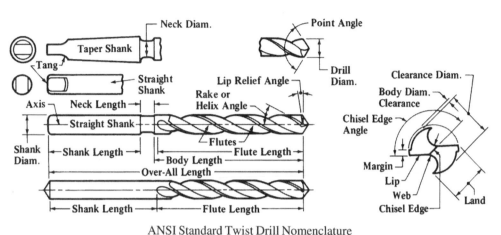

ANSI Standard Twist Drill Nomenclature

Figure 5.7 Parts of a twist drill

Figure 5.7 summarizes the parts of a twist drill. The diameter of the drill is measured across the margin of the drill. Larger diameter drills are driven by their tapered shank. The point angle for general purpose drills is 118 degrees.

The size of the drill is stamped or imprinted on the shank of the drill. Drill sizes are expressed in one of the following ways:

- Number   #97 (.0059) to # 1 (.228)
- Letter   A (.234) to Z (.413)
- Fractional   1/64 to 1.00 and larger
- Metric   mm

Shop Recommended Straight Shank Twist Drills pages 868–875, Table 1

868   TWIST DRILLS

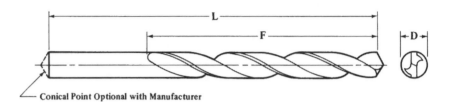

Conical Point Optional with Manufacturer

Table 1. ANSI Straight Shank Twist Drills — Jobbers Length through 17.5 mm, Taper Length through 12.7 mm, and Screw Machine Length through 25.4 mm Diameter *ANSI/ASME B94.11M-1993*

> Drill sizes begin with the smallest diameter expressed by letters of the alphabet, beginning with #97 which is .0059. Note that the fractional system overlaps the alphabet system at 1/64″ or .0156

Drill Diameter, D^a				Jobbers Length				Taper Length				Screw Machine Length			
Frac-tion No. or Ltr.		Equivalent		Flute		Overall		Flute		Overall		Flute		Overall	
		Decimal		F		L		F		L		F		L	
	mm	In.	mm	Inch	mm	Inch	mm	Inch	mm	Inch	mm	Inch	mm	Inch	mm
97	0.15	0.0059	0.150	1/16	1.6	3/4	19	…	…	…	…	…	…	…	…
96	0.16	0.0063	0.160	1/16	1.6	3/4	19	…	…	…	…	…	…	…	…
95	0.17	0.0067	0.170	1/16	1.6	3/4	19	…	…	…	…	…	…	…	…
94	0.18	0.0071	0.180	1/16	1.6	3/4	19	…	…	…	…	…	…	…	…
93	0.19	0.0075	0.190	1/16	1.6	3/4	19	…	…	…	…	…	…	…	…
92	0.20	0.0079	0.200	1/16	1.6	3/4	19	…	…	…	…	…	…	…	…
91		0.0083	0.211	5/64	2.0	3/4	19	…	…	…	…	…	…	…	…
90	0.22	0.0087	0.221	5/64	2.0	3/4	19	…	…	…	…	…	…	…	…
89		0.0091	0.231	5/64	2.0	3/4	19	…	…	…	…	…	…	…	…
88		0.0095	0.241	5/64	2.0	3/4	19	…	…	…	…	…	…	…	…
	0.25	0.0098	0.250	5/64	2.0	3/4	19	…	…	…	…	…	…	…	…
87		0.0100	0.254	5/64	2.0	3/4	19	…	…	…	…	…	…	…	…
86		0.0105	0.267	3/32	2.4	3/4	19	…	…	…	…	…	…	…	…
85	0.28	0.0110	0.280	3/32	2.4	3/4	19	…	…	…	…	…	…	…	…
84		0.0115	0.292	3/32	2.4	3/4	19	…	…	…	…	…	…	…	…
	0.30	0.0118	0.300	3/32	2.4	3/4	19	…	…	…	…	…	…	…	…
83		0.0120	0.305	3/32	2.4	3/4	19	…	…	…	…	…	…	…	…
82		0.0125	0.318	3/32	2.4	3/4	19	…	…	…	…	…	…	…	…
	0.32	0.0126	0.320	3/32	2.4	3/4	19	…	…	…	…	…	…	…	…
81		0.0130	0.330	3/32	2.4	3/4	19	…	…	…	…	…	…	…	…
80		0.0135	0.343	1/8	3	3/4	19	…	…	…	…	…	…	…	…
	0.35	0.0138	0.350	1/8	3	3/4	19	…	…	…	…	…	…	…	…
79		0.0145	0.368	1/8	3	3/4	19	…	…	…	…	…	…	…	…
	0.38	0.0150	0.380	3/16	5	3/4	19	…	…	…	…	…	…	…	…
1/64		0.0156	0.396	3/16	5	3/4	19	…	…	…	…	…	…	…	…
	0.40	0.0157	0.400	3/16	5	3/4	19	…	…	…	…	…	…	…	…
78		0.0160	0.406	3/16	5	7/8	22	…	…	…	…	…	…	…	…
	0.42	0.0165	0.420	3/16	5	7/8	22	…	…	…	…	…	…	…	…
	0.45	0.0177	0.450	3/16	5	7/8	22	…	…	…	…	…	…	…	…
77		0.0180	0.457	3/16	5	7/8	22	…	…	…	…	…	…	…	…
	0.48	0.0189	0.480	3/16	5	7/8	22	…	…	…	…	…	…	…	…
	0.50	0.0197	0.500	3/16	5	7/8	22	…	…	…	…	…	…	…	…
76		0.0200	0.508	3/16	5	7/8	22	…	…	…	…	…	…	…	…
75		0.0210	0.533	1/4	6	1	25	…	…	…	…	…	…	…	…
	0.55	0.0217	0.550	1/4	6	1	25	…	…	…	…	…	…	…	…
74		0.0225	0.572	1/4	6	1	25	…	…	…	…	…	…	…	…
	0.60	0.0236	0.600	5/16	8	1 1/8	29	…	…	…	…	…	…	…	…

Table 1. *(Continued)* **ANSI Straight Shank Twist Drills — Jobbers Length through 17.5 mm, Taper Length through 12.7 mm, and Screw Machine Length through 25.4 mm Diameter** *ANSI/ASME B94.11M-1993*

Frac-tion No. or Ltr.	mm	Decimal In.	mm	Flute F Inch	Flute F mm	Overall L Inch	Overall L mm	Flute F Inch	Flute F mm	Overall L Inch	Overall L mm	Flute F Inch	Flute F mm	Overall L Inch	Overall L mm
	5.10	0.2008	5.100	2 7/16	62	3 5/8	92	3 5/8	92	6	152	1 3/16	30	2 1/4	57
7		0.2010	5.105	2 7/16	62	3 5/8	92	3 5/8	92	6	152	1 3/16	30	2 1/4	57
13/64		0.2031	5.159	2 7/16	62	3 5/8	92	3 5/8	92	6	152	1 3/16	30	2 1/4	57
6		0.2040	5.182	2 1/2	64	3 3/4	95	3 5/8	92	6	152	1 1/4	32	2 3/8	60
	5.20	0.2047	5.200	2 1/2	64	3 3/4	95	3 5/8	92	6	152	1 1/4	32	2 3/8	60
5		0.2055	5.220	2 1/2	64	3 3/4	95	3 5/8	92	6	152	1 1/4	32	2 3/8	60
	5.30	0.2087	5.300	2 1/2	64	3 3/4	95	3 5/8	92	6	152	1 1/4	32	2 3/8	60
4		0.2090	5.309	2 1/2	64	3 3/4	95	3 5/8	92	6	152	1 1/4	32	2 3/8	60
	5.40	0.2126	5.400	2 1/2	64	3 3/4	95	3 5/8	92	6	152	1 1/4	32	2 3/8	60
3		0.2130	5.410	2 1/2	64	3 3/4	95	3 5/8	92	6	152	1 1/4	32	2 3/8	60
	5.50	0.2165	5.500	2 1/2	64	3 3/4	95	3 5/8	92	6	152	1 1/4	32	2 3/8	60
7/32		0.2188	5.558	2 1/2	64	3 3/4	95	3 5/8	92	6	152	1 1/4	32	2 3/8	60
	5.60	0.2205	5.600	2 5/8	67	3 7/8	98	3 3/4	95	6 1/8	156	1 5/16	33	2 7/16	62
2		0.2210	5.613	2 5/8	67	3 7/8	98	3 3/4	95	6 1/8	156	1 5/16	33	2 7/16	62
	5.70	0.2244	5.700	2 5/8	67	3 7/8	98	3 3/4	95	6 1/8	156	1 5/16	33	2 7/16	62
1		0.2280	5.791	2 5/8	67	3 7/8	98	3 3/4	95	6 1/8	156	1 5/16	33	2 7/16	62
	5.80	0.2283	5.800	2 5/8	67	3 7/8	98	3 3/4	95	6 1/8	156	1 5/16	33	2 7/16	62
	5.90	0.2323	5.900	2 5/8	67	3 7/8	98	3 3/4	95	6 1/8	156	1 5/16	33	2 7/16	62
A		0.2340	5.944	2 5/8	67	3 7/8	98	…	…	…	…	1 5/16	33	2 7/16	62
15/64		0.2344	5.954	2 5/8	67	3 7/8	98	3 3/4	95	6 1/8	156	1 5/16	33	2 7/16	62
	6.00	0.2362	6.000	2 3/4	70	4	102	3 3/4	95	6 1/8	156	1 3/8	35	2 1/2	64
B		0.2380	6.045	2 3/4	70	4	102	…	…	…	…	1 3/8	35	2 1/2	64
	6.10	0.2402	6.100	2 3/4	70	4	102	3 3/4	95	6 1/8	156	1 3/8	35	2 1/2	64
C		0.2420	6.147	2 3/4	70	4	102	…	…	…	…	1 3/8	35	2 1/2	64
	6.20	0.2441	6.200	2 3/4	70	4	102	3 3/4	95	6 1/8	156	1 3/8	35	2 1/2	64
D		0.2460	6.248	2 3/4	70	4	102	…	…	…	…	1 3/8	35	2 1/2	64
	6.30	0.2480	6.300	2 3/4	70	4	102	3 3/4	95	6 1/8	156	1 3/8	35	2 1/2	64
E, 1/4		0.2500	6.350	2 3/4	70	4	102	3 3/4	95	6 1/8	156	1 3/8	35	2 1/2	64
	6.40	0.2520	6.400	2 7/8	73	4 1/8	105	3 7/8	98	6 1/4	159	1 7/16	37	2 5/8	67
	6.50	0.2559	6.500	2 7/8	73	4 1/8	105	3 7/8	98	6 1/4	159	1 7/16	37	2 5/8	67
F		0.2570	6.528	2 7/8	73	4 1/8	105	…	…	…	…	1 7/16	37	2 5/8	67
	6.60	0.2598	6.600	2 7/8	73	4 1/8	105	…	…	…	…	1 7/16	37	2 5/8	67
G		0.2610	6.629	2 7/8	73	4 1/8	105	…	…	…	…	1 7/16	37	2 5/8	67
	6.70	0.2638	6.700	2 7/8	73	4 1/8	105	…	…	…	…	1 7/16	37	2 5/8	67
17/64		0.2656	6.746	2 7/8	73	4 1/8	105	3 7/8	98	6 1/4	159	1 7/16	37	2 5/8	67
H		0.2660	6.756	2 7/8	73	4 1/8	105	…	…	…	…	1 1/2	38	2 11/16	68
	6.80	0.2677	6.800	2 7/8	73	4 1/8	105	3 7/8	98	6 1/4	159	1 1/2	38	2 11/16	68
	6.90	0.2717	6.900	2 7/8	73	4 1/8	105	…	…	…	…	1 1/2	38	2 11/16	68
I		0.2720	6.909	2 7/8	73	4 1/8	105	…	…	…	…	1 1/2	38	2 11/16	68
	7.00	0.2756	7.000	2 7/8	73	4 1/8	105	3 7/8	98	6 1/4	159	1 1/2	38	2 11/16	68
J		0.2770	7.036	2 7/8	73	4 1/8	105	…	…	…	…	1 1/2	38	2 11/16	68
	7.10	0.2795	7.100	2 15/16	75	4 1/4	108	…	…	…	…	1 1/2	38	2 11/16	68
K		0.2810	7.137	2 15/16	75	4 1/4	108	…	…	…	…	1 1/2	38	2 11/16	68
9/32		0.2812	7.142	2 15/16	75	4 1/4	108	3 7/8	98	6 1/4	159	1 1/2	38	2 11/16	68
	7.20	0.2835	7.200	2 15/16	75	4 1/4	108	4	102	6 3/8	162	1 9/16	40	2 3/4	70
	7.30	0.2874	7.300	2 15/16	75	4 1/4	108	…	…	…	…	1 9/16	40	2 3/4	70

The alphabet system begins with letter A at .234. With three systems of drill sizes, there is a drill diameter available every few thousandths of an inch among the smaller sizes.

5

5

Table 1. *(Continued)* **ANSI Straight Shank Twist Drills — Jobbers Length through 17.5 mm, Taper Length through 12.7 mm, and Screw Machine Length through 25.4 mm Diameter** *ANSI/ASME B94.11M-1993*

Drill Diameter, D[a]				Jobbers Length				Taper Length				Screw Machine Length			
Frac-tion No. or Ltr.	mm	Equivalent		Flute		Overall		Flute		Overall		Flute		Overall	
		Decimal In.	mm	F		L		F		L		F		L	
				Inch	mm	Inch	mm	Inch	mm	Inch	mm	Inch	mm	Inch	mm
X		0.3970	10.084	3¾	95	5⅛	130	...	...	...	...	1¹⁵⁄₁₆	49	3⁵⁄₁₆	84
	10.20	0.4016	10.200	3⅞	98	5¼	133	4⅜	111	7	178	1¹⁵⁄₁₆	49	3⁵⁄₁₆	84
Y		0.4040	10.262	3⅞	98	5¼	133	...	...	...	...	1¹⁵⁄₁₆	49	3⁵⁄₁₆	84
¹³⁄₃₂		0.4062	10.317	3⅞	98	5¼	133	4⅜	111	7	178	1¹⁵⁄₁₆	49	3⁵⁄₁₆	84
Z		0.4130	10.490	3⅞	98	5¼	133	...	...	...	...	2	51	3⅜	86
	10.50	0.4134	10.500	3⅞	98	5¼	133	4⅝	117	7¼	184	2	51	3⅜	86
²⁷⁄₆₄		0.4219	10.716	3¹⁵⁄₁₆	100	5⅜	137	4⅝	117	7¼	184	2	51	3⅜	86
	10.80	0.4252	10.800	4¹⁄₁₆	103	5½	140	4⅝	117	7¼	184	2¹⁄₁₆	52	3⁷⁄₁₆	87
	11.00	0.4331	11.000	4¹⁄₁₆	103	5½	140	4⅝	117	7¼	184	2¹⁄₁₆	52	3⁷⁄₁₆	87
⁷⁄₁₆		0.4375	11.112	4¹⁄₁₆	103	5½	140	4⅝	117	7¼	184	2¹⁄₁₆	52	3⁷⁄₁₆	87
	11.20	0.4409	11.200	4³⁄₁₆	106	5⅝	143	4¾	121	7½	190	2⅛	54	3⁹⁄₁₆	90
	11.50	0.4528	11.500	4³⁄₁₆	106	5⅝	143	4¾	121	7½	190	2⅛	54	3⁹⁄₁₆	90
²⁹⁄₆₄		0.4531	11.509	4³⁄₁₆	106	5⅝	143	4¾	121	7½	190	2⅛	54	3⁹⁄₁₆	90
	11.80	0.4646	11.800	4⁵⁄₁₆	110	5¾	146	4¾	121	7½	190	2⅛	54	3⅝	92
¹⁵⁄₃₂		0.4688	11.908	4⁵⁄₁₆	110	5¾	146	4¾	121	7½	190	2⅛	54	3⅝	92
	12.00	0.4724	12.000	4⅜	111	5⅞	149	4¾	121	7¾	197	2³⁄₁₆	56	3¹¹⁄₁₆	94
	12.20	0.4803	12.200	4⅜	111	5⅞	149	4¾	121	7¾	197	2³⁄₁₆	56	3¹¹⁄₁₆	94
³¹⁄₆₄		0.4844	12.304	4⅜	111	5⅞	149	4¾	121	7¾	197	2³⁄₁₆	56	3¹¹⁄₁₆	94
	12.50	0.4921	12.500	4½	114	6	152	4¾	121	7¾	197	2¼	57	3¾	95
½		0.5000	12.700	4½	114	6	152	4¾	121	7¾	197	2¼	57	3¾	95
	12.80	0.5039	12.800	4½	114	6	152	...	...	...	...	2⅜	60	3⅞	98
	13.00	0.5118	13.000	4½	114	6	152	...	...	...	...	2⅜	60	3⅞	98
³³⁄₆₄		0.5156	13.096	4¹³⁄₁₆	122	6⅝	168	...	...	...	...	2⅜	60	3⅞	98
	13.20	0.5197	13.200	4¹³⁄₁₆	122	6⅝	168	...	...	...	...	2⅜	60	3⅞	98
¹⁷⁄₃₂		0.5312	13.492	4¹³⁄₁₆	122	6⅝	168	...	...	...	...	2⅜	60	3⅞	98
	13.50	0.5315	13.500	4¹³⁄₁₆	122	6⅝	168	...	...	...	...	2⅜	60	3⅞	98
	13.80	0.5433	13.800	4¹³⁄₁₆	122	6⅝	168	...	...	...	...	2½	64	4	102
³⁵⁄₆₄		0.5469	13.891	4¹³⁄₁₆	122	6⅝	168	...	...	...	...	2½	64	4	102
	14.00	0.5512	14.000	4¹³⁄₁₆	122	6⅝	168	...	...	...	...	2½	64	4	102
	14.25	0.5610	14.250	4¹³⁄₁₆	122	6⅝	168	...	...	...	...	2½	64	4	102
⁹⁄₁₆		0.5625	14.288	4¹³⁄₁₆	122	6⅝	168	...	...	...	...	2½	64	4	102
	14.50	0.5709	14.500	4¹³⁄₁₆	122	6⅝	168	...	...	...	...	2⅝	67	4⅛	105
³⁷⁄₆₄		0.5781	14.684	4¹³⁄₁₆	122	6⅝	168	...	...	...	...	2⅝	67	4⅛	105
	14.75	0.5807	14.750	5³⁄₁₆	132	7⅛	181	...	...	...	...	2⅝	67	4⅛	105
	15.00	0.5906	15.000	5³⁄₁₆	132	7⅛	181	...	...	...	...	2⅝	67	4⅛	105
¹⁹⁄₃₂		0.5938	15.083	5³⁄₁₆	132	7⅛	181	...	...	...	...	2⅝	67	4⅛	105
	15.25	0.6004	15.250	5³⁄₁₆	132	7⅛	181	...	...	...	...	2¾	70	4¼	108
³⁹⁄₆₄		0.6094	15.479	5³⁄₁₆	132	7⅛	181	...	...	...	...	2¾	70	4¼	108
	15.50	0.6102	15.500	5³⁄₁₆	132	7⅛	181	...	...	...	...	2¾	70	4¼	108
	15.75	0.6201	15.750	5³⁄₁₆	132	7⅛	181	...	...	...	...	2¾	70	4¼	108
⁵⁄₈		0.6250	15.875	5³⁄₁₆	132	7⅛	181	...	...	...	...	2¾	70	4¼	108
	16.00	0.6299	16.000	5³⁄₁₆	132	7⅛	181	...	...	...	...	2⅞	73	4½	114
	16.25	0.6398	16.250	5³⁄₁₆	132	7⅛	181	...	...	...	...	2⅞	73	4½	114
⁴¹⁄₆₄		0.6406	16.271	5³⁄₁₆	132	7⅛	181	...	...	...	...	2⅞	73	4½	144
	16.50	0.6496	16.500	5³⁄₁₆	132	7⅛	181	...	...	...	...	2⅞	73	4½	114
²¹⁄₃₂		0.6562	16.669	5³⁄₁₆	132	7⅛	181	...	...	...	...	2⅞	73	4½	114

Drills larger than 1/2″ diameter are usually held by a tapered shank because the cutting force can make a straight shank drill slip.

Unit G: Taps pages 904–947

Producing threaded holes with a thread cutting tool is called *tapping*. Taps have two, three, or four cutting flutes. Taps either are driven by hand with a tap wrench or are machine driven (see Figure 5.8). Cutting oil is used when tapping most materials. Cast iron, bronze, and brass are tapped dry or tapped with oil that has special properties. When hand tapping, the tap wrench is reversed periodically to break the chip. When machine tapping, the tap is driven to depth, the spindle is reversed, and the tap is withdrawn.

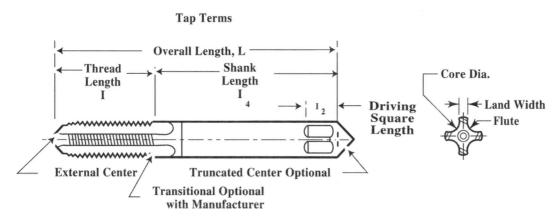

Figure 5.8 Typical tap

Drilling the correct size hole before tapping is crucial. If the hole is too small, tapping effort will be increased, often resulting in a broken tap. If the hole is too large, there will be an inadequate amount of thread engagement. Tap drill sizes are based on about 75% effective thread. A full thread is only about 5% stronger than a 75% thread and the tapping effort is much greater. The depth of useable threads in a threaded hole should be from *one and a half to two times the screw diameter*. Tapping holes deeper than two times the diameter of the thread is not advised because the threads will strip before the head of the screw breaks off.

For a 3/4 — 10 thread, the effective thread depth is 1.125″ to 1.50″ (2 × 3/4 = 1.5). Figure 5.9 shows a common thread designation.

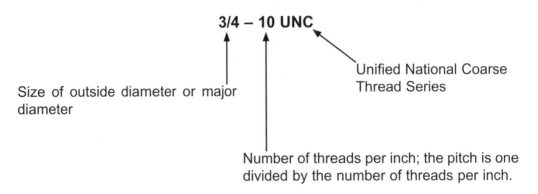

Figure 5.9 Common thread designation

Metric threads are expressed differently (see Figure 5.10):

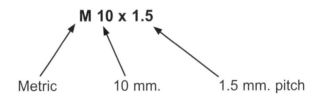

Figure 5.10 Metric threads

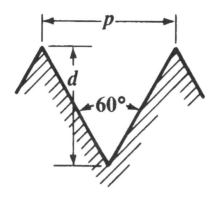

Figure 5.11 Pitch

The *pitch* of a thread (P) is one divided by the number of threads per inch (see Figure 5.11).

The tap drill diameter is found by subtracting the pitch from the major diameter. On metric threads, this calculation is easy: 10 minus 1.5 = 8.5 mm. This method can be used for general purpose threads. It is useful when the thread you are producing does not appear on a chart such as Table 2 on pages 2021–2028.

Example
The tap drill for a 3/4–10 tap is: 3/4 minus 1/10 or

$$
\begin{array}{r}
.750 \\
-\,.100 \\
\hline
.650
\end{array}
$$

Shop Recommended page 2030, Table 4

Table 4. Tap Drills and Clearance Drills for Machine Screws with American National Thread Form

Size of Screw		No. of Threads per Inch	Tap Drills		Clearance Hole Drills			
No. or Diam.	Decimal Equiv.		Drill Size	Decimal Equiv.	Close Fit		Free Fit	
					Drill Size	Decimal Equiv.	Drill Size	Decimal Equiv.
0	.060	80	$\frac{3}{64}$	.0469	52	.0635	50	.0700
1	.073	64	53	.0595	48	.0760	46	.0810
		72	53	.0595				
2	.086	56	50	.0700	43	.0890	41	.0960
		64	50	.0700				
3	.099	48	47	.0785	37	.1040	35	.1100
		56	45	.0820				
4	.112	36[a]	44	.0860	32	.1160	30	.1285
		40	43	.0890				
		48	42	.0935				
5	.125	40	38	.1015	30	.1285	29	.1360
		44	37	1040				
6	.138	32	36	.1065	27	.1440	25	.1495
		40	33	.1130				
8	.164	32	29	.1360	18	.1695	16	.1770
		36	29	.1360				
10	.190	24	25	.1495	9	.1960	7	.2010
		32	21	.1590				
12	.216	24	16	.1770	2	.2210	1	.2280
		28	14	.1820				
14	.242	20[a]	10	.1935	D	.2460	F	.2570
		24[a]	7	.2010				
$\frac{1}{4}$	.250	20	7	.2010	F	.2570	H	.2660
		28	3	.2130				
$\frac{5}{16}$	.3125	18	F	.2570	P	.3230	Q	.3320
		24	I	.2720				
$\frac{3}{8}$	.375	16	$\frac{5}{16}$	.3125	W	.3860	X	.3970
		24	Q	.3320				
$\frac{7}{16}$	.4375	14	U	.3680	$\frac{29}{64}$	.4531	$\frac{15}{32}$	.4687
		20	$\frac{25}{64}$	.3906				
$\frac{1}{2}$	.500	13	$\frac{27}{64}$	.4219	$\frac{33}{64}$	.5156	$\frac{17}{32}$	.5312
		20	$\frac{29}{64}$	.4531				

Tooling and Toolmaking

Shop Recommended page 1808, Figure 1

Figure 1 describes terms and characteristics of a common thread.

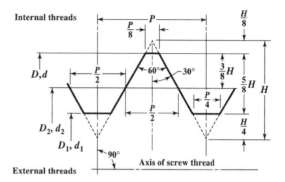

On an external thread, the *major diameter* is the size of the outside of the thread, also called the *crest.* The *minor diameter* is the bottom of the thread, also called the *root.*

On an internal thread, the *major diameter* is the largest diameter of the thread and the *minor diameter* is determined by the tap drill size. There are advantages to drilling through holes for tapping.

- The depth of effective thread is easier to obtain and chips produced by the tapping operation are free to drop out the bottom of the hole.
- Through holes can also be used to transfer locations to other components.
- Broken taps are easier to remove from a through hole.

Holes that do not go through the material are *blind holes*. The depth of a blind hole for a tap must include about 6 times the *pitch* of the screw. This is to allow for the first few threads of the tap which are a lead for the full thread.

The major causes of broken taps are:

- Impacted chips or bottoming out in *blind holes*
- Starting the tap crooked
- Worn tap

Analyze, Evaluate, & Implement

Choose four common tap sizes. List the proper drilled depth for each size. Allow for:

- 1 1/2 times the thread diameter
- 6 times the pitch plus .06 for end clearance

SECTION 5

A file is a tool that is used for smoothing or deburring sharp edges left from a previous machining operation. Files are classified according to their cross-sectional shape or the spacing and type of teeth. Like hacksaws, machinists' files cut only on the forward stroke. "Saw Files" are for sharpening saws (see "Figure 1: Styles of Mill Files").

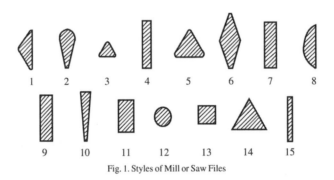

Fig. 1. Styles of Mill or Saw Files

1. Cantsaw File
2. Crosscut File
3. Double Ender File
4. Mill File
5. Triangular Saw File or Taper Saw File
6. Web Saw File
7. Flat File
8. Half-Round File

9. Hand File
10. Knife File
11. Pillar File
12. Round File
13. Square File
14. Three Square File
15. Warding File

Of the different classes of files, the types used the most by machinists and toolmakers are the:

- Machinist's file
- Swiss Pattern file

Machinist's files are for general purpose and are usually flat or slightly rounded on one side. They may be single cut (having one row of teeth) or double cut (having two rows of diagonally crossing teeth). Machinist's files include the mill file which is single cut and may be used for draw filing. Draw filing is done by holding the file perpendicular to the work and carefully pushing the file away from you and dragging it back. This method results in a smooth uniform cut.

Swiss pattern files are made in the same shapes as machinist's files but they are smaller and thinner and are made to closer tolerances. Swiss pattern files may be used by skilled toolmakers to blend surfaces, to finish delicate objects, or to fit precise sections.

Burs

Burs are used in an electric or air-operated rotary device to smooth or shape areas that are difficult to reach with other methods. Most burs used in the machine shop are 1/8 to 1/4 in diameter. Carbide burs operated by a high speed, air-operated grinder are common.

Unit L: Tool Wear and Sharpening pages 996–1003

This unit copies selected excerpts from *Machinery's Handbook 29*, followed by a bulleted list of simplified explanations.

Shop Recommended page 996 Definitions

TOOL WEAR AND SHARPENING

Metal cutting tools wear constantly when they are being used. A normal amount of wear should not be a cause for concern until the size of the worn region has reached the point where the tool should be replaced. Normal wear cannot be avoided and should be differentiated from abnormal tool breakage or excessively fast wear. Tool breakage and an excessive rate of wear indicate that the tool is not operating correctly and steps should be taken to correct this situation.

Making It Simple

- The cutting tool must be harder than the workpiece.
- All cutting tools wear, the process of predicting wear is the key.

Shop Recommended Cratering page 997

Cratering.—A deep crater will sometimes form on the face of the tool which is easily recognizable. The crater forms at a short distance behind the side cutting edge leaving a small shelf between the cutting edge and the edge of the crater. This shelf is sometimes covered with the built-up edge and at other times it is uncovered. Often the bottom of the crater is obscured with work material that is welded to the tool in this region. Under normal operating conditions, the crater will gradually enlarge until it breaks through a part of the cutting edge. Usually this occurs on the end cutting edge just behind the nose. When this takes place, the flank wear at the nose increases rapidly and complete tool failure follows shortly. Sometimes cratering cannot be avoided and a slow increase in the size of the crater is considered normal. However, if the rate of crater growth is rapid, leading to a short tool life, corrective measures must be taken.

5

Making It Simple

■ Cratering is the tendency of a material to weld itself to the cutting tool, break off, and cause the tool to chip in a crater-like shape.

Shop Recommended page 997 Cutting Edge Chipping

Cutting Edge Chipping.—Small chips are sometimes broken from the cutting edge which accelerates tool wear but does not necessarily cause immediate tool failure. Chipping can be recognized by the appearance of the cutting edge and the flank wear land. A sharp depression in the lower edge of the wear land is a sign of chipping and if this edge of the wear land has a jagged appearance it indicates that a large amount of chipping has taken place. Often the vacancy or cleft in the cutting edge that results from chipping is filled up with work material that is tightly welded in place. This occurs very rapidly when chipping is caused by a built-up edge on the face of the tool. In this manner the damage to the cutting edge is healed; however, the width of the wear land below the chip is usually increased and the tool life is shortened.

Making it Simple

■ Friction welding can occur between the workpiece and the cutting tool.
■ This condition is common and may be regarded as normal.
■ This condition may escalate and cause premature wear.

Shop Recommended page 997

Deformation.—Deformation occurs on carbide cutting tools when taking a very heavy cut using a slow cutting speed and a high feed rate. A large section of the cutting edge then becomes very hot and the heavy cutting pressure compresses the nose of the cutting edge, thereby lowering the face of the tool in the area of the nose. This reduces the relief under the nose, increases the width of the wear land in this region, and shortens the tool life.

Making It Simple

■ Using proper speeds and feeds is the basis for long cutting tool life.
■ Depth of cut has an important role in cutting tool life.

🔍 **For more Information on *Speeds and Feeds*, see Section 6, Machining Operations, Unit B.**

Shop Recommended page 998 Surface Finish

Surface Finish.—The finish on the machined surface does not necessarily indicate poor cutting tool performance unless there is a rapid deterioration. A good surface finish is, however, sometimes a requirement. The principal cause of a poor surface finish is the built-up edge which forms along the edge of the cutting tool. The elimination of the built-up edge will always result in an improvement of the surface finish. The most effective way to eliminate the built-up edge is to increase the cutting speed. When the cutting speed is increased beyond a certain critical cutting speed, there will be a rather sudden and large improvement in the surface finish. Cemented carbide tools can operate successfully at higher cutting speeds, where the built-up edge does not occur and where a good surface finish is obtained. Whenever possible, cemented carbide tools should be operated at cutting speeds where a good surface finish will result. There are times when such speeds are not possible. Also, high-speed tools cannot be operated at the speed where the built-up edge does not form. In these conditions the most effective method of obtaining a good surface finish is to employ a cutting fluid that has active sulphur or chlorine additives.

5

Making It Simple

- ■ Proper speeds and feeds directly influence surface finish.
- ■ Higher spindle speeds produce good surface finishes.
- ■ The use of cutting fluid improves surface finish.

🔍 **For more information on *cutting fluids*, see Section 6, Unit F Machining Operations, *Machinery's Handbook Made* Easy, pages 141–142.**

🔍 **Index navigation paths and key words:**

— cutting fluids/for different materials/page 1184

SECTION 5

Shop Recommended page 998 Sharpening Twist Drills

Sharpening Twist Drills.—Twist drills are cutting tools designed to perform concurrently several functions, such as penetrating directly into solid material, ejecting the removed chips outside the cutting area, maintaining the essentially straight direction of the advance movement and controlling the size of the drilled hole. The geometry needed for these multiple functions is incorporated into the design of the twist drill in such a manner that it can be retained even after repeated sharpening operations. Twist drills are resharpened many times during their service life, with the practically complete restitution of their original operational characteristics. However, in order to assure all the benefits which the design of the twist drill is capable of providing, the surfaces generated in the sharpening process must agree with the original form of the tool's operating surfaces, unless a change of shape is required for use on a different work material.

Making It Simple

- Twist drills can be sharpened repeatedly if the angles and clearances are maintained.
- Drills may be sharpened by hand on a pedestal grinder.
- The angles of the tool can be changed to drill materials that have different properties.

Shop Recommended page 998–999

The principal elements of the tool geometry which are essential for the adequate cutting performance of twist drills are shown in Fig. 1. The generally used values for these dimensions are the following:

Point angle: Commonly 118°, except for high strength steels, 118° to 135°; aluminum alloys, 90° to 140°; and magnesium alloys, 70° to 118°.

Helix angle: Commonly 24° to 32°, except for magnesium and copper alloys, 10° to 30°.

Lip relief angle: Commonly 10° to 15°, except for high strength or tough steels, 7° to 12°. The lower values of these angle ranges are used for drills of larger diameter, the higher values for the smaller diameters. For drills of diameters less than $\frac{1}{4}$ inch (6.35 mm), the lip relief angles are increased beyond the listed maximum values up to 24°. For soft and free machining materials, 12° to 18° except for diameters less than $\frac{1}{4}$ inch (6.35 mm), 20° to 26°.

Shop Recommended page 999 Relief Grinding

The relief grinding of the flank surfaces will generate the chisel angle on the web of the twist drill. The value of that angle, typically 55°, which can be measured, for example, with the protractor of an optical projector, is indicative of the correctness of the relief grinding.

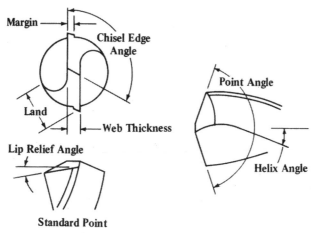

Fig. 1. The principal elements of tool geometry on twist drills.

Making It Simple

- Most drills for metal have a point angle of 118°.
- The *point angle* for harder materials is flatter (larger) than for soft materials.
- The *helix angle* is the spiral channel of the drill.
- The *lip relief angle* does not do the cutting; it is required for strength and clearance.

5

Shop Recommended page 999 Grinding the Face of a Twist Drill

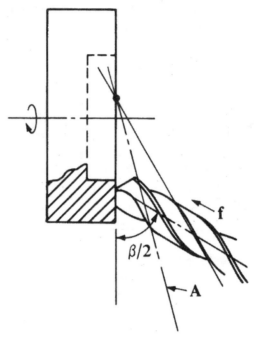

Fig. 2. In grinding the face of the twist drill the tool is swung aroud the axis *A*
of an imaginary cone, while resting in a support tilted by half of the point
angle *β* with respect to the face of the grinding wheel. Feed *f* for
stock removal is in the direction of the drill axis.

Shop Recommended page 999–1000

Drill Point Thinning.—The chisel edge is the least efficient operating surface element of
the twist drill because it does not cut, but actually squeezes or extrudes the work material.
To improve the inefficient cutting conditions caused by the chisel edge, the point width is
often reduced in a drill-point thinning operation, resulting in a condition such as that shown
in Fig. 3. Point thinning is particularly desirable on larger size drills and also on those
which become shorter in usage, because the thickness of the web increases toward the shaft
of the twist drill, thereby adding to the length of the chisel edge. The extent of point thinning is limited by the minimum strength of the web needed to avoid splitting of the drill
point under the influence of cutting forces.

Shop Recommended page 1000 Sharpening operations

Both sharpening operations—the relieved face grinding and the point thinning—should be carried out in special drill grinding machines or with twist drill grinding fixtures mounted on general-purpose tool grinding machines, designed to assure the essential accuracy of the required tool geometry. Off-hand grinding may be used for the important web thinning when a special machine is not available; however, such operation requires skill and experience.

Improperly sharpened twist drills, e.g. those with unequal edge length or asymmetrical point angle, will tend to produce holes with poor diameter and directional control.

For deep holes and also drilling into stainless steel, titanium alloys, high temperature alloys, nickel alloys, very high strength materials and in some cases tool steels, split point grinding, resulting in a "crankshaft" type drill point, is recommended. In this type of pointing, see Fig. 4, the chisel edge is entirely eliminated, extending the positive rake cutting edges to the center of the drill, thereby greatly reducing the required thrust in drilling. Points on modified-point drills must be restored after sharpening to maintain their increased drilling efficiency.

5

Making It Simple

- As a drill gets shorter from sharpening, the web increases.
- The increased web of a drill makes it hard for the drill to penetrate the material.
- Web thinning can be done by hand, but special tool grinders are recommended.
- The geometry of the drill angles can be changed to suit metals with different properties.

Unit C: Forming Tools pages 796–807

Unit D: Milling Cutters pages 808–843

Forming Tools reproduce their shape in the workpiece. The contour of the cutting edge corresponds to the shape required. Forming Tools are used in turning machines and milling machines. One of the most common forming tools is an angle cutter used on a milling machine to mill a surface at an angle.

Index navigation paths and key words:

Milling Cutters are available in the solid type with various numbers of cutting teeth and the inserted blade type which holds a number of replaceable inserts. Very often, more than one milling cutter is used to machine the required shape, one example being the dovetail cutter. In this operation, a straight end mill is used to rough out a slot before the angled cut is made.

Index navigation paths and key words:

Unit H: Standard Tapers pages 948–971

The tools and machines of industry use tapered shafts to align machine components and drive cutting tools. Producing tapered shafts is a common machine shop task. Check Unit H for a description of the most popular types of standard tapers, their uses, dimensions, and characteristics.

Index navigation paths and key words:

Unit I: Arbors, Chucks, and Spindles pages 972–978

Machine spindles, arbors, and chucks drive components of special machinery, grinding wheels, cutters, and saw blades.

Index navigation paths and key words:

Unit J: Broaches and Broaching pages 979–985

Broaching is a process that is applied to internal features such as holes in cases where irregular shapes are required. Broaches are used to produce keyways in bores of machine parts.

Index navigation paths and key words:

— broaches/types/pages 979–980
— broaching /difficulties/page 985
— broaching/cutting oils for/page 1185

ASSIGNMENT

List the key terms and give a definition of each.

Positive Rake	Shank	Pitch
Negative Rake	Point Angle	Major Diameter
Nose Radius	Margin	Minor Diameter
High Speed Steel	Tapered Shank	Blind Hole
Reaming Allowance	Flute	Cratering
Helical Flute	Effective Thread	

APPLY IT! PART 1

1. How many cutting edges are on a negative rake, square insert? A triangular insert?

2. Refer to page 741 in the Machinery's Handbook: Identify the features of the following insert:

$$T \quad N \quad L \quad A \quad 4 \quad 3 \quad 2 \qquad B$$

3. Which insert shape is the best choice for:

   a. A heavy interrupted cut?

   b. Machining the flat end (facing) of a 4.00" diameter shaft?

   c. Finish machining a 2.500" diameter hole (boring) with a flat bottom inside of a 3.75" diameter shaft?

   d. A deep groove in a shaft?

4. A shaft machined at 250 RPM with a high speed steel cutter can be turned at about_____ RPM using a carbide cutting tool.

5. How are carbide inserts held in the tool holder?

6. The carbide insert that is the most versatile is the _____.

SECTION 5

APPLY IT! PART 2

Fill in the blanks in the table below.

Reamer Size	.2500	.3125	.3750
Drill #1			
Drill #2			

APPLY IT! PART 3

1. What is the decimal equivalent of a 3/8 diameter drill?

2. Your engineering drawing calls for three .147 drilled holes. What size drill should be used?

3. What part of the drill is held by the machine tool?

4. What is the size range for the Alphabet drill size system?

5. The fractional drill size system begins at _____

6. What is the metric equivalent of a Y drill?

APPLY IT! PART 4

1. Calculate the proper tap drill for an M 6 × 1.0. Convert the result to inches.

2. The letter system of drill sizes goes from _____ to _____

3. The major diameter of a 3/8 — 16 screw is _____. The pitch is _____.

4. How do metric thread designations differ from the inch system?

5. To have adequate holding power, how much thread engagement should these screws have?
   a. 1/2–13
   b. M 8
   c. 3/8–16

6. What size tap drill should be used for the following thread sizes?
   a. 1/4–20
   b. 5/16–18
   c. 3/8–24
   d. 3/8–16
   e. 1/2–20

Section **6**

MACHINING OPERATIONS

Use this section with pages 1004 — 1327
in Machinery's Handbook 29th Edition

SECTION 6

Navigation Overview

Units Covered in this Section

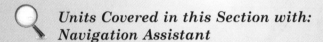

Units Covered in this Section with:
Navigation Assistant

The Navigation Assistant helps find information in the MH29 Primary Index. The Primary Index is located in the back of the book on pages 2701-2788 and is set up alphabetically by subject. Watch for the magnifying glass throughout the section for navigation hints.

May I help you?

Key Terms

High Speed Steel	Plain Carbon Steel
RPM	Ferrous and Non-Ferrous
Tool Steel	Alloy Steel
Cutting Speed	Surface Feet per Minute
Thermal Shock	Cratering

Machining Operations

Learning Objectives

After studying this unit you should be able to:

- **State the benefits and properties of high speed steel and carbide cutters.**
- **Use the tables in MHB to set speeds and feeds on a variety of machine tools.**
- **Explain how cutting speed relates to material properties.**
- **Create a coordinate list.**
- **Plot points using the Cartesian Coordinate System.**

Introduction

6

Most materials that are machined are metals. Different metals have different machining characteristics. With a few exceptions, softer metals are easier to machine than harder metals. The definition of *hardness* is "resistance to penetration." A metal's hardness is tested on a machine called a *hardness tester* and reported using different scales such as *Brinell* and *Rockwell*.

The spindle speeds of most machine tools are adjustable to accommodate the different characteristics of different materials. Spindle speeds are expressed as RPM or *revolutions per minute*.

For example, to drill a hole using a drill press, the RPM of the drill press is selected based on the *cutting speed* of the material, the drill material, the diameter of the drill, and the capability of the machine tool. Tables 1 through 14 on pages 1026–1034 (in Machinery Handbook) show the *cutting speeds* of different metals. The cutting speed is the number of feet of material that pass the cutting tool in one minute. The terms *cs* (cutting speed), *sfm* (surface feet per minute), and *fpm* (feet per minute) are interchangeable terms. Softer metals that machine easily, such as aluminum, have higher cutting speeds than harder metals.

Unit A: Cutting Speeds and Feeds pages 1008–1020

Unit B: Speed and Feed Tables pages 1021–1080

SECTION 6

High Speed Steel

 Index navigation path and key words:

— steel/high speed/page 438

For machining to take place, the cutting tool must be harder than the workpiece. Two commonly used cutting tool materials are *high speed steel* (HSS) and *carbide*. High speed steels are divided into two major groups based on their alloying elements:

- Tungsten type
- Molybdenum type

The principal properties of these steels are the ability to withstand high temperatures and abrasion and to retain their hardness deeply into the cutting tool. Most high speed steel cutting tools are used in turning machines such as lathes. The tool may assume many different shapes based on the machining operation. Unlike other harder cutting tools, high speed steel cutting tools can be easily shaped on a pedestal grinder. High speed steel cutting tools are not usually used for production or for machining tough, high alloy steels.

Carbide

 Index navigation path and key terms:

— carbide tools/pages 779, 785

Tough, abrasive materials can be machined much more easily with carbide tools. Higher spindle speeds are required, which means greater productivity. There are hundreds of different grades of carbide, but there are four distinct types: *straight tungsten carbide, crater-resistant carbides, titanium carbides, and coated carbides.*

Straight Tungsten Carbides
Straight tungsten carbides work well on gray cast iron, non-metals, and non-ferrous (not containing iron) metals.

Crater-Resistant Carbides
Crater-resistant carbides work well on steel, alloy steel, and materials that tend to *crater. Cratering* is the tendency of a material to weld itself to the cutting tool, break off, and cause the tool to chip in a crater-like shape.

Titanium Carbides
Titanium carbides work well at high spindle speeds using light cuts. For this reason, they are used for "hard turning," which is machining hardened material at high speeds. They work best for continuous-chip machining where there is no interruption in the cut.

Coated Carbides

Coated carbides are used for general machining of steel and alloy steel. They are tough and resistant to thermal shock and will tolerate interrupted cuts. Thermal shock is the rapid heating and cooling of the cutting tool, causing it to crack.

Safety First

- **Chips produced by carbide cutters are hot and razor sharp. Chip shields are available to contain the chips and keep them away from the operator.**

Speeds and Feeds

 Index navigation path and key terms:

— cutting speeds and feeds/pages 1008–1073

The cutting speed of a material refers to its machineability. For example, the machineability of mild steel is expressed as its *cutting speed*, which is about 100 when using high speed steel cutters. The terms *cutting speed* and *surface feet per minute* are interchangeable. Using this as point of reference, materials that machine easier than mild steel have higher cutting speeds and materials that do not machine as easily as mild steel have lower cutting speeds. Aluminum, for example, machines very easily and has a higher cutting speed of steel. High cutting speeds mean high rpms at the machine spindle.

Feeds on turning machines are expressed in *inches per revolution (ipr)*. On a lathe, when the spindle makes one revolution the cutter advances the feed selected.

Feeds on milling machines are expressed in *inches per minute (ipm)*. Unlike turning machines, the feed on milling machines is set independently of spindle RPM.

Machine tools can be set up using ideal speeds and feeds, but there are other factors such as depth of cut, rigidity, available horsepower, and the use of coolant. The skilled machinist must balance these variables to maximize productivity and work in a safe manner.

Selecting the Cutting Conditions

Shop Recommended page 1013

Selecting Cutting Conditions.—The first step in establishing cutting conditions is to select depth of cut. The depth of cut will be limited by the amount of metal to be machined from the workpiece, by the power available on the machine tool, by the rigidity of the workpiece and cutting tool, and by the rigidity of the setup. Depth of cut has the least effect upon tool life, so the heaviest possible depth of cut should always be used.

The second step is to select the feed (feed/rev for turning, drilling, and reaming, or feed/tooth for milling). The available power must be sufficient to make the required depth of cut at the selected feed. The maximum feed possible that will produce an acceptable surface finish should be selected.

The third step is to select the cutting speed. Although the accompanying tables provide recommended cutting speeds and feeds for many materials, experience in machining a certain material may form the best basis for adjusting given cutting speeds to a particular job. In general, depth of cut should be selected first, followed by feed, and last cutting speed.

In order to solve the cutting speed and feed formulas, a basic understanding of algebra is required. Algebra is the branch of mathematics that uses variables such as numbers to solve equations. Algebra is reviewed in *Machinery's Handbook* 29 in Mathematics.

 Index navigation path and key words:

— algebra and equations /page 30

Shop Recommended Cutting Speed Formulas page 1015

Cutting Speed Formulas

Most machining operations are conducted on machine tools having a rotating spindle. Cutting speeds are usually given in feet or meters per minute and these speeds must be converted to spindle speeds, in revolutions per minute, to operate the machine. Conversion is accomplished by use of the following formulas:

For U.S. units:

$$N = \frac{12V}{\pi D} = \frac{12 \times 252}{\pi \times 8} = 120 \text{ rpm}$$

For metric units:

$$N = \frac{1000V}{\pi D} = 318.3 \frac{V}{D} \text{ rpm}$$

where N is the spindle speed in revolutions per minute (rpm); V is the cutting speed in feet per minute (fpm) for U.S. units and meters per minute (m/min) for metric units. In turning, D is the diameter of the workpiece; in milling, drilling, reaming, and other operations that use a rotating tool, D is the cutter diameter in inches for U.S. units and in millimeters for metric units. $\pi = 3.1416$.

Machining Operations

The cutting speed formulas are reproduced here. The spindle speed of the machine is set using the *feet per minute* (*fpm*) found on Tables 1–17 on pages 997–1033 and the formula that follows.

 Index navigation path and key words:

— cutting speeds and feeds/formulas for/pages 1015, 1035

For example, if the material is 3.0 diameter 1020 mild steel being machined on a lathe with high speed steel, the formula looks like this:

Constant (part of the formula)

Cutting speed (*fpm*) from Table 1, page 997

$$N(rpm) = \frac{12(120)}{\pi(3.0)}$$

3.1416

or

$$N(rpm) = \frac{1440}{9.4248}$$

$$N(rpm) = 152.78$$

Making It Simple

To simplify the process of RPM calculations, the following formula is an accepted substitute:

$$RPM = \frac{4 \times cutting\ speed}{diameter}$$

Enter the information from the previous example:

$$RPM = \frac{4 \times 120}{3.0}$$

$$RPM = 160$$

Almost the same answer, but less work. Determining an RPM is a balance between productivity, tool wear, and safety. Calculated RPMs are just a starting point. The most efficient machine setting is a process that is based on many variables such as rigidity, available horsepower, depth of cut, and the use of coolant.

6

SECTION 6

Example

In the next example, the machine is a vertical milling machine. The cutter is a carbide face mill; 6.0 inch diameter with 8 carbide inserts. The engineering drawing identifies the material as 4140 alloy steel with a Brinell Hardness of 230. The cutting speed found on page 1046; Table 11 is 200 (optimum) to 320 (average). The average cutting speed is chosen with a chipload or "feed per tooth" of 20 (0.02). The diameter of the *cutter* is entered in this case.

$$RPM = \frac{4 \times 320}{6.0}$$

$$RPM = 213.3$$

The feed for milling machines is expressed in "inches per minute" (ipm). The formula for determining milling feed is:

Feed in inches per minute = RPM × feed per tooth × number of teeth on the cutter

Enter known values:

213.3 × .020 inches per tooth × 8 inserts = 34.128 inches per minute

---

Analyze, Evaluate, & Implement

- How does the simplified version of the RPM formula differ from the cutting speed formulas from MHB 29; page 1015?
- Make a list of the alloying elements and their percentages in the tungsten type and molybdenum type of high speed steels.

Machining Operations

Shop Recommended page 1016

Cutting Speeds and Equivalent RPM for Drills of Number and Letter Sizes

> This chart is for drilling. Find the drill size in the left hand column and find the corresponding cutting speed from the top of the chart.

> For fractional drill sizes, see pages 1017–1018.

> Example:
> The material is 1020 steel. From Table 17 on page 1060, a cutting speed of 90 is shown. The drill size is "U" or .368. The RPM is 934.

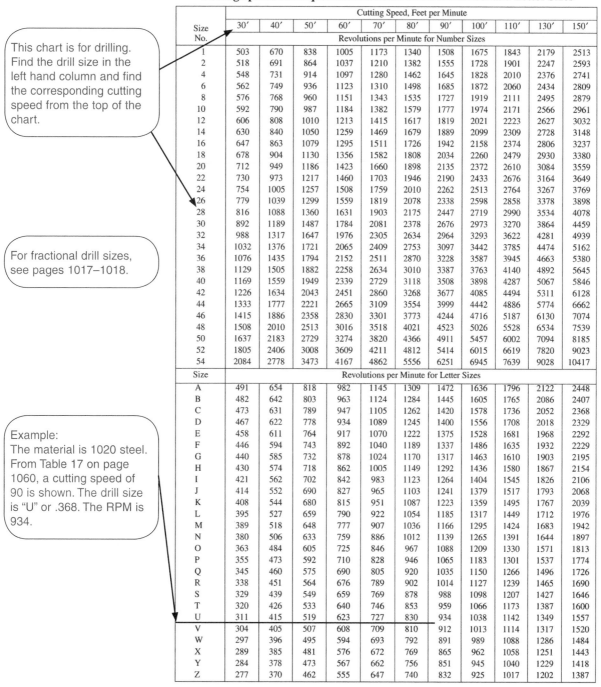

Size No.	Cutting Speed, Feet per Minute										
	30′	40′	50′	60′	70′	80′	90′	100′	110′	130′	150′
	Revolutions per Minute for Number Sizes										
1	503	670	838	1005	1173	1340	1508	1675	1843	2179	2513
2	518	691	864	1037	1210	1382	1555	1728	1901	2247	2593
4	548	731	914	1097	1280	1462	1645	1828	2010	2376	2741
6	562	749	936	1123	1310	1498	1685	1872	2060	2434	2809
8	576	768	960	1151	1343	1535	1727	1919	2111	2495	2879
10	592	790	987	1184	1382	1579	1777	1974	2171	2566	2961
12	606	808	1010	1213	1415	1617	1819	2021	2223	2627	3032
14	630	840	1050	1259	1469	1679	1889	2099	2309	2728	3148
16	647	863	1079	1295	1511	1726	1942	2158	2374	2806	3237
18	678	904	1130	1356	1582	1808	2034	2260	2479	2930	3380
20	712	949	1186	1423	1660	1898	2135	2372	2610	3084	3559
22	730	973	1217	1460	1703	1946	2190	2433	2676	3164	3649
24	754	1005	1257	1508	1759	2010	2262	2513	2764	3267	3769
26	779	1039	1299	1559	1819	2078	2338	2598	2858	3378	3898
28	816	1088	1360	1631	1903	2175	2447	2719	2990	3534	4078
30	892	1189	1487	1784	2081	2378	2676	2973	3270	3864	4459
32	988	1317	1647	1976	2305	2634	2964	3293	3622	4281	4939
34	1032	1376	1721	2065	2409	2753	3097	3442	3785	4474	5162
36	1076	1435	1794	2152	2511	2870	3228	3587	3945	4663	5380
38	1129	1505	1882	2258	2634	3010	3387	3763	4140	4892	5645
40	1169	1559	1949	2339	2729	3118	3508	3898	4287	5067	5846
42	1226	1634	2043	2451	2860	3268	3677	4085	4494	5311	6128
44	1333	1777	2221	2665	3109	3554	3999	4442	4886	5774	6662
46	1415	1886	2358	2830	3301	3773	4244	4716	5187	6130	7074
48	1508	2010	2513	3016	3518	4021	4523	5026	5528	6534	7539
50	1637	2183	2729	3274	3820	4366	4911	5457	6002	7094	8185
52	1805	2406	3008	3609	4211	4812	5414	6015	6619	7820	9023
54	2084	2778	3473	4167	4862	5556	6251	6945	7639	9028	10417
Size	Revolutions per Minute for Letter Sizes										
A	491	654	818	982	1145	1309	1472	1636	1796	2122	2448
B	482	642	803	963	1124	1284	1445	1605	1765	2086	2407
C	473	631	789	947	1105	1262	1420	1578	1736	2052	2368
D	467	622	778	934	1089	1245	1400	1556	1708	2018	2329
E	458	611	764	917	1070	1222	1375	1528	1681	1968	2292
F	446	594	743	892	1040	1189	1337	1486	1635	1932	2229
G	440	585	732	878	1024	1170	1317	1463	1610	1903	2195
H	430	574	718	862	1005	1149	1292	1436	1580	1867	2154
I	421	562	702	842	983	1123	1264	1404	1545	1826	2106
J	414	552	690	827	965	1103	1241	1379	1517	1793	2068
K	408	544	680	815	951	1087	1223	1359	1495	1767	2039
L	395	527	659	790	922	1054	1185	1317	1449	1712	1976
M	389	518	648	777	907	1036	1166	1295	1424	1683	1942
N	380	506	633	759	886	1012	1139	1265	1391	1644	1897
O	363	484	605	725	846	967	1088	1209	1330	1571	1813
P	355	473	592	710	828	946	1065	1183	1301	1537	1774
Q	345	460	575	690	805	920	1035	1150	1266	1496	1726
R	338	451	564	676	789	902	1014	1127	1239	1465	1690
S	329	439	549	659	769	878	988	1098	1207	1427	1646
T	320	426	533	640	746	853	959	1066	1173	1387	1600
U	311	415	519	623	727	830	934	1038	1142	1349	1557
V	304	405	507	608	709	810	912	1013	1114	1317	1520
W	297	396	495	594	693	792	891	989	1088	1286	1484
X	289	385	481	576	672	769	865	962	1058	1251	1443
Y	284	378	473	567	662	756	851	945	1040	1229	1418
Z	277	370	462	555	647	740	832	925	1017	1202	1387

For fractional drill sizes, use the following table.

This table continues on page 1017, showing larger fractional sizes.

6

SECTION 6

How to Use the Feeds and Speeds Table

A list of all the speed and feed tables is found on page 1021. Use this information to find the correct speed and feed table for the machine tool and material you are working with.

Shop Recommended page 1021

Principal Speed and Feed Tables

Feeds and Speeds for Turning
Table 1. Cutting Feeds and Speeds for Turning Plain Carbon and Alloy Steels
Table 2. Cutting Feeds and Speeds for Turning Tool Steels
Table 3. Cutting Feeds and Speeds for Turning Stainless Steels
Table 4a. Cutting Feeds and Speeds for Turning Ferrous Cast Metals
Table 4b. Cutting Feeds and Speeds for Turning Ferrous Cast Metals
Table 5c. Cutting-Speed Adjustment Factors for Turning with HSS Tools
Table 5a. Turning-Speed Adjustment Factors for Feed, Depth of Cut, and Lead Angle
Table 5b. Tool Life Factors for Turning with Carbides, Ceramics, Cermets, CBN, and Polycrystalline Diamond
Table 6. Cutting Feeds and Speeds for Turning Copper Alloys
Table 7. Cutting Feeds and Speeds for Turning Titanium and Titanium Alloys
Table 8. Cutting Feeds and Speeds for Turning Light Metals
Table 9. Cutting Feeds and Speeds for Turning Superalloys

Feeds and Speeds for Milling
Table 10. Cutting Feeds and Speeds for Milling Aluminum Alloys
Table 11. Cutting Feeds and Speeds for Milling Plain Carbon and Alloy Steels
Table 12. Cutting Feeds and Speeds for Milling Tool Steels
Table 13. Cutting Feeds and Speeds for Milling Stainless Steels
Table 14. Cutting Feeds and Speeds for Milling Ferrous Cast Metals
Table 15a. Recommended Feed in Inches per Tooth (ft) for Milling with High Speed Steel Cutters
Table 15b. End Milling (Full Slot) Speed Adjustment Factors for Feed, Depth of Cut, and Lead Angle
Table 15c. End, Slit, and Side Milling Speed Adjustment Factors for Radial Depth of Cut
Table 15d. Face Milling Speed Adjustment Factors for Feed, Depth of Cut, and Lead Angle
Table 15e. Tool Life Adjustment Factors for Face Milling, End Milling, Drilling, and Reaming
Table 16. Cutting Tool Grade Descriptions and Common Vendor Equivalents

Feeds and Speeds for Drilling, Reaming, and Threading
Table 17. Feeds and Speeds for Drilling, Reaming, and Threading Plain Carbon and Alloy Steels
Table 18. Feeds and Speeds for Drilling, Reaming, and Threading Tool Steels
Table 19. Feeds and Speeds for Drilling, Reaming, and Threading Stainless Steels
Table 20. Feeds and Speeds for Drilling, Reaming, and Threading Ferrous Cast Metals
Table 21. Feeds and Speeds for Drilling, Reaming, and Threading Light Metals
Table 22. Feed and Diameter Speed Adjustment Factors for HSS Twist Drills and Reamers
Table 23. Feeds and Speeds for Drilling and Reaming Copper Alloys

Tables 1–9 are for turning machines such as lathes and turning centers.

Tables 10–16 are for milling machines such as "Bridgeports," vertical milling machines, computer numerical control (CNC) milling machines, and horizontal spindle milling machines.

Tables 17–23 are for drilling, reaming, and threading regardless of the type of machine tool that is used for these operations. All of the tables are grouped by similar machining characteristics.

6

Machining Operations

Shop Recommended page 1022 Combined feed/speed portion of the tables

The combined feed/speed portion of the speed tables gives two sets of feed and speed data for each material represented. These feed/speed pairs are the *optimum* and *average* data (identified by *Opt.* and *Avg.*); the *optimum* set is always on the left side of the column and the *average* set is on the right. The *optimum* feed/speed data are approximate values of feed and speed that achieve minimum-cost machining by combining a high productivity rate with low tooling cost at a fixed tool life. The *average* feed/speed data are expected to achieve approximately the same tool life and tooling costs, but productivity is usually lower, so machining costs are higher. The data in this portion of the tables are given in the form of two numbers, of which the first is the feed in thousandths of an inch per revolution (or per tooth, for milling) and the second is the cutting speed in feet per minute. For example, the feed/speed set 15/215 represents a feed of 0.015 in/rev (0.38 mm/rev) at a speed of 215 fpm (65.6 m/min). Blank cells in the data tables indicate that feed/speed data for these materials were not available at the time of publication.

Shop Recommended Example 1, Turning pages 1025, 1027

6

Example 1, Turning: Find the cutting speed for turning SAE 1074 plain carbon steel of 225 to 275 Brinell hardness, using an uncoated carbide insert, a feed of 0.015 in./rev, and a depth of cut of 0.1 inch.

In Table 1, feed and speed data for two types of uncoated carbide tools are given, one for hard tool grades, the other for tough tool grades. In general, use the speed data from the tool category that most closely matches the tool to be used because there are often significant differences in the speeds and feeds for different tool grades. From the uncoated carbide hard grade values, the *optimum* and *average* feed/speed data given in Table 1 are 17/615 and 8/815, or 0.017 in./rev at 615 ft/min and 0.008 in./rev at 815 ft/min. Because the selected feed (0.015 in./rev) is different from either of the feeds given in the table, the cutting speed must be adjusted to match the feed. The other cutting parameters to be used must also be compared with the general tool and cutting parameters given in the speed tables to determine if adjustments need to be made for these parameters as well. The general tool and cutting parameters for turning, given in the footnote to Table 1, are depth of cut = 0.1 inch, lead angle = 15°, and tool nose radius = $\frac{3}{64}$ inch.

Safety First

- **Long, stringy chips produced by turning operations are razor sharp and dangerous. Increasing the feed rate or changing the cutting tool geometry can make the chips break off and be more manageable.**

Shop Recommended page 1026, 1027; Table 1

If you are working from an engineering drawing, the type of material will be given in the title block or in a general note.

> The cutting tool material is a critical factor in determining the surface feet per minute (sfm).

Table 1. Cutting Feeds and Speeds for Turning Plain Carbon and Alloy Steels

Material AISI/SAE Designation	Brinell Hardness	HSS Speed (fpm)		Uncoated Carbide Hard Opt.	Avg.	Tough Opt.	Avg.	Coated Carbide Hard Opt.	Avg.	Tough Opt.	Avg.	Ceramic Hard Opt.	Avg.	Tough Opt.	Avg.	Cermet Opt.	Avg.
Free-machining plain carbon steels (resulfurized): 1212, 1213, 1215	100-150	150	f / s	17 / 805	8 / 1075	36 / 405	17 / 555	17 / 1165	8 / 1295	28 / 850	13 / 1200	15 / 3340	8 / 4985	15 / 1670	8 / 2500	7 / 1610	3 / 2055
	150-200	160	f / s	17 / 745	8 / 935	36 / 345	17 / 470	28 / 915	13 / 1130	28 / 785	13 / 1110	15 / 1795	8 / 2680	15 / 1485	8 / 2215	7 / 1490	3 / 1815
1108, 1109, 1115, 1117, 1118, 1120, 1126, 1211	100-150	130															
	150-200	120	f / s	17 / 730	8 / 990	36 / 300	17 / 430	17 / 1090	8 / 1410	28 / 780	13 / 1105	15 / 1610	8 / 2780	15 / 1345	8 / 2005	7 / 1355	3 / 1695
1132, 1137, 1139, 1140, 1144, 1146, 1151	175-225	120	f / s	17 / 615	8 / 815	36 / 300	17 / 405	17 / 865	8 / 960	28 / 755	13 / 960	13 / 1400	7 / 1965	13 / 1170	7 / 1640		
	275-325	75															
	325-375	50	f / s	17 / 515	8 / 685	36 / 235	17 / 340	17 / 720	8 / 805	28 / 650	13 / 810	10 / 1430	5 / 1745	10 / 1070	5 / 1305		
	375-425	40															
(Leaded): 11L17, 11L18, 12L13, 12L14	100-150	140															
	150-200	145	f / s	17 / 745	8 / 935	36 / 345	17 / 470	28 / 915	13 / 1130	28 / 785	13 / 1110	15 / 1795	8 / 2680	15 / 1485	8 / 2215	7 / 1490	3 / 1815
	200-250	110	f / s	17 / 615	8 / 815	36 / 300	17 / 405	17 / 865	8 / 960	28 / 755	13 / 960	13 / 1400	7 / 1965	13 / 1170	7 / 1640		
Plain carbon steels: 1006, 1008, 1009, 1010, 1012, 1015, 1016, 1017, 1018, 1019, 1020, 1021, 1022, 1023, 1024, 1025, 1026, 1513, 1514	100-125	120	f / s	17 / 805	8 / 1075	36 / 405	17 / 555	17 / 1165	8 / 1295	28 / 850	13 / 1200	15 / 3340	8 / 4985	15 / 1670	8 / 2500	7 / 1610	3 / 2055
	125-175	110	f / s	17 / 745	8 / 935	36 / 345	17 / 470	28 / 915	13 / 1130	28 / 785	13 / 1110	15 / 1795	8 / 2680	15 / 1485	8 / 2215	7 / 1490	3 / 1815
	175-225	90	f / s	17 / 615	8 / 815	36 / 300	17 / 405	17 / 865	8 / 960	28 / 755	13 / 960	13 / 1400	7 / 1965	13 / 1170	7 / 1640		
	225-275	70	f / s														

f = feed (0.001 in./rev), s = speed (ft/min) Metric Units: f × 25.4 = mm/rev, s × 0.3048 = m/min

> The *Brinell Hardness(Bhn)* is a measurement of the materials' resistance to penetration and is useful for determining machineability.

> Uncoated and coated carbide cutting tools require higher cutting speeds.

> The use of ceramic and cermet cutters is limited to machine tools with the horsepower and high spindle speeds required for these inserts.

> High Speed Steel (HSS) cutters can be easily shaped by hand on a pedestal grinder. Speeds for HSS tools are based on a feed of 0.012 per revolution and a depth of cut of 0.125.

Table 1. *(Continued)* **Cutting Feeds and Speeds for Turning Plain Carbon and Alloy Steels**

Material AISI/SAE Designation	Brinell Hardness	HSS Speed (fpm)		Uncoated Carbide					Coated Carbide					Ceramic					Cermet	
				Hard		Tough			Hard		Tough			Hard		Tough				
				Opt.	Avg.	Opt.	Avg.		Opt.	Avg.	Opt.	Avg.		Opt.	Avg.	Opt.	Avg.		Opt.	Avg.
Plain carbon steels *(continued)*: 1027, 1030, 1033, 1035, 1036, 1037, 1038, 1039, 1040, 1041, 1042, 1043, 1045, 1046, 1048, 1049, 1050, 1052, 1524, 1526, 1527, 1541	125-175	100	f	17	8	36	17		28	13	28	13		15	8	15	8		7	3
	175-225	85	s	745	935	345	470		915	1130	785	1110		1795	2680	1485	2215		1490	1815
	225-275	70	f	17	8	36	17		17	8	28	13		13	7	13	7			
	275-325	60	s	615	815	300	405		865	960	755	960		1400	1965	1170	1640			
	325-375	40	f	17	8	36	17		17	8	28	13		10	5	10	5			
	375-425	30	s	515	685	235	340		720	805	650	810		1430	1745	1070	1305			
Plain carbon steels *(continued)*: 1055, 1060, 1064, 1065, 1070, 1074, 1078, 1080, 1084, 1086, 1090, 1095, 1548, 1551, 1552, 1561, 1566	125-175	100	f	17	8	36	17		17	8	28	13		15	8	15	8		7	3
	175-225	80	s	730	990	300	430		1090	1410	780	1105		1610	2780	1345	2005		1355	1695
	225-275	65	f	17	8	36	17		17	8	28	13		13	7	13	7		7	3
	275-325	50	s	615	815	300	405		865	960	755	960		1400	1965	1170	1640		1365	1695
	325-375	35	f	17	8	36	17		17	8	28	13		10	5	10	5			
	375-425	30	s	515	685	235	340		720	805	650	810		1430	1745	1070	1305			
Free-machining alloy steels, (resulfurized): 4140, 4150	175-200	110	f	17	8	36	17		17	8	28	13		15	8	15	8		7	3
	200-250	90	s	525	705	235	320		505	525	685	960		1490	2220	1190	1780		1040	1310
	250-300	65	f	17	8	36	17		17	8	28	13		10	5	10	5		7	3
	300-375	50	s	355	445	140	200		630	850	455	650		1230	1510	990	1210		715	915
	375-425	40	f		8	36	17		17	8	28	13		8	4	8	4		7	3
			s	330	440	125	175		585	790	125	220		1200	1320	960	1060		575	740

Tool Material. **f** = feed (0.001 in./rev), **s** = speed (ft/min)   *Metric Units:* **f** × 25.4 = mm/rev, **s** × 0.3048 = m/min

"f" is the feed rate in inches per revolution (ipr) in thousandths. For example, "17" is .017; "8" is .008. Every time the spindle of the turning machine makes one revolution, the cutting tool advances the feed rate selected on the machine. The table gives two sets of speed and feed data for each material listed. These feed/speed pairs are the optimum and average (identified by *Opt* and *Avg*).

"s" is the surface feet per minute of material that passes the point of the cutting tool in one minute. Use this value to select the proper spindle speed in RPM.

6

Analyze, Evaluate, & Implement

Using the formula,

$$RPM = \frac{4 \times \textit{cutting speed}}{\textit{diameter}}$$

determine the rpm for turning a 2 1/2 diameter shaft with a high speed steel (HSS) cutter. The material is SAE 1035 steel with a Brinell Hardness of 230. The cutting speed is 70.

Entering the information we know, and using Table 1 from page 1027,

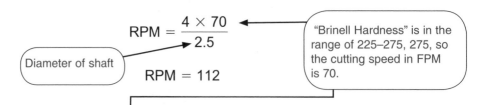

$$RPM = \frac{4 \times 70}{2.5}$$

Diameter of shaft

"Brinell Hardness" is in the range of 225–275, 275, so the cutting speed in FPM is 70.

$$RPM = 112$$

Table 1. *(Continued)* **Cutting Feeds and Speeds for Turning Plain Carbon and Alloy Steels**

Material AISI/SAE Designation	Brinell Hardness	HSS Speed (fpm)		Uncoated Carbide				Coated Carbide				Ceramic				Cermet	
				Hard		Tough		Hard		Tough		Hard		Tough			
							f = feed (0.001 in./rev), s = speed (ft/min) Metric Units: f × 25.4 = mm/rev, s × 0.3048 = m/min										
				Opt.	Avg.	Opt.	Avg.	Opt.	Avg.	Opt.	Avg.	Opt.	Avg.	Opt.	Avg.	Opt.	Avg.
Plain carbon steels *(continued)*: 1027, 1030, 1033, 1035, 1036, 1037, 1038, 1039, 1040, 1041, 1042, 1043, 1045, 1046, 1048, 1049, 1050, 1052, 1524, 1526, 1527, 1541	125-175	100	f	17	8	36	17	28	13	28	13	15	8	15	8	7	3
			s	745	935	345	470	915	1130	785	1110	1795	2680	1485	2215	1490	1815
	175-225	85															
	225-275	70	f	17	8	36	17	17	8	28	13	13	7	13	7		
			s	615	815	300	405	865	960	755	960	1400	1965	1170	1640		
	275-325	60															
	325-375	40	f	17	8	36	17	17	8	28	13	10	5	10	5		
			s	515	685	235	340	720	805	650	810	1430	1745	1070	1305		
	375-425	30															
Plain carbon steels *(continued)*: 1055, 1060, 1064, 1065, 1070, 1074, 1078, 1080, 1084, 1086, 1090, 1095, 1548, 1551, 1552, 1561, 1566	125-175	100															
	175-225	80	f	17	8	36	17	17	8	28	13	15	8	15	8	7	3
			s	730	990	300	430	1090	1410	780	1105	1610	2780	1345	2005	1355	1695
	225-275	65	f	17	8	36	17	17	8	28	13	13	7	13	7	7	3
			s	615	815	300	405	865	960	755	960	1400	1965	1170	1640	1365	1695
	275-325	50															
	325-375	35	f	17	8	36	17	17	8	28	13	10	5	10	5		
			s	515	685	235	340	720	805	650	810	1430	1745	1070	1305		
	375-425	30															
Free-machining alloy steels, (resulfurized): 4140, 4150	175-200	110															
	200-250	90	f	17	8	36	17	17	8	28	13	15	8	15	8	7	3
			s	525	705	235	320	505	525	685	960	1490	2220	1190	1780	1040	1310
	250-300	65	f	17	8	36	17	17	8	28	13	10	5	10	5	7	3
			s	355	445	140	200	630	850	455	650	1230	1510	990	1210	715	915
	300-375	50															
	375-425	40	f	17	8	36	17	17	8	28	13	8	4	8	4	7	3
			s	330	440	125	175	585	790	125	220	1200	1320	960	1060	575	740

Shop Recommended page 1044, Table 11

Table 11, page 1044, shows cutting speeds and feeds for milling machines such as Bridgeport vertical milling machines using End Mills, Face Mills, and Slit Mills

End mill

Face Mill

Slit Mill

Machining Operations

Table 11. Cutting Feeds and Speeds for Milling Plain Carbon and Alloy Steels

f = feed (0.001 in./tooth), s = speed (ft/min) *Metric Units:* f × 25.4 = mm/rev, s × 0.3048 = m/min

Material	Brinell Hardness	HSS Speed (fpm)	f/s	End Milling HSS Opt.	End Milling HSS Avg.	End Milling Uncoated Carbide Opt.	End Milling Uncoated Carbide Avg.	End Milling Coated Carbide Opt.	End Milling Coated Carbide Avg.	Face Milling Uncoated Carbide Opt.	Face Milling Uncoated Carbide Avg.	Face Milling Coated Carbide Opt.	Face Milling Coated Carbide Avg.	Slit Milling Uncoated Carbide Opt.	Slit Milling Uncoated Carbide Avg.	Slit Milling Coated Carbide Opt.	Slit Milling Coated Carbide Avg.
Free-machining plain carbon steels (resulfurized): 1212, 1213, 1215	100-150	140	f	7	4	7	4	7	4	39	20	39	20	39	20	39	20
			s	45	125	465	735	800	1050	225	335	415	685	265	495	525	830
	150-200	130	f	7	4							39	20				
			s	35	100							215	405				
(Resulfurized): 1108, 1109, 1115, 1117, 1118, 1120, 1126, 1211	100-150	130	f	7	4	7	4	7	4	39	20	39	20	39	20	39	20
	150-200	115	s	730	85	325	565	465	720	140	220	195	365	170	350	245	495
	175-225	115	f	7	4							39	20				
			s	30	85							185	350				
(Resulfurized): 1132, 1137, 1139, 1140, 1144, 1146, 1151	275-325	70															
	325-375	45	f	7	4	7	4	7	4	39	20	39	20	39	20	39	20
			s	25	70	210	435	300	560	90	170	175	330	90	235	135	325
	375-425	35															
(Leaded): 11L17, 11L18, 12L13, 12L14	100-150	140	f	7	4							39	20				
	150-200	130	s	35	100							215	405				
	200-250	110	f	7	4							39	20				
			s	30	85							185	350				
Plain carbon steels: 1006, 1008, 1009, 1010, 1012, 1015, 1016, 1017, 1018, 1019, 1020, 1021, 1022, 1023, 1024, 1025, 1026, 1513, 1514	100-125	110	f	7	4	7	4	7	4	39	20	39	20	39	20	39	20
			s	45	125	465	735	800	1050	225	335	415	685	265	495	525	830
	125-175	110	f	7	4							39	20				
			s	35	100							215	405				
	175-225	90	f	7	4							39	20				
	225-275	65	s	30	85							185	350				

> Most steel used in manufacturing is plain carbon steel, also known as cold rolled steel or mild steel.

> One factor of the milling feed equation is the feed per tooth, or chip load, shown here as a whole number. The value is given in thousandths; "7" is .007 and "4" is .004.

Analyze, Evaluate, & Implement

Use the Feeds and Speeds tables on pages 1044–1047 to set RPM and feed.

6

135

Table 11. Cutting Feeds and Speeds for Milling Plain Carbon and Alloy Steels

			End Milling						Face Milling				Slit Milling			
		HSS	HSS		Uncoated Carbide		Coated Carbide		Uncoated Carbide		Coated Carbide		Uncoated Carbide		Coated Carbide	
Material	Brinell Hardness	Speed (fpm)	f = feed (0.001 in./tooth), s = speed (ft/min) *Metric Units:* f × 25.4 = mm/rev, s × 0.3048 = m/min													
			Opt.	Avg.	Opt.	Avg.	Opt.	Avg.	Opt.	Avg.	Opt.	Avg.	Opt.	Avg.	Opt.	Avg.
Free-machining plain carbon steels (resulfurized): 1212, 1213, 1215	100-150	140	f 7 / s 45	4 / 125	7 / 465	4 / 735	7 / 800	4 / 1050	39 / 225	20 / 335	39 / 415	20 / 685	39 / 265	20 / 495	39 / 525	20 / 830
	150-200	130	f 7 / s 35	4 / 100							39 / 215	20 / 405				
(Resulfurized): 1108, 1109, 1115, 1117, 1118, 1120, 1126, 1211	100-150	130	f / s 730	4 / 85	7 / 325	4 / 565	7 / 465	4 / 720	39 / 140	20 / 220	39 / 195	20 / 365	39 / 170	20 / 350	39 / 245	20 / 495
	150-200	115														
	175-225	115	f 7 / s 30	4 / 85									39 / 185	20 / 350		
(Resulfurized): 1132, 1137, 1139, 1140, 1144, 1146, 1151	275-325	70														
	325-375	45	f 7 / s 25	4 / 70	7 / 210	4 / 435	7 / 300	4 / 560	39 / 90	20 / 170	39 / 175	20 / 330	39 / 90	20 / 235	39 / 135	20 / 325
	375-425	35														
(Leaded): 11L17, 11L18, 12L13, 12L14	100-150	140	f 7 / s 35	4 / 100							39 / 215	20 / 405				
	150-200	130														
	200-250	110	f 7 / s 30	4 / 85							39 / 185	20 / 350				
Plain carbon steels: 1006, 1008, 1009, 1010, 1012, 1015, 1016, 1017, 1018, 1019, 1020, 1021, 1022, 1023, 1024, 1025, 1026, 1513, 1514	100-125	110	f 7 / s 45	4 / 125	7 / 465	4 / 735	7 / 800	4 / 1050	39 / 225	20 / 335	39 / 415	20 / 685	39 / 265	20 / 495	39 / 525	20 / 830
	125-175	110	f 7 / s 35	4 / 100							39 / 215	20 / 405				
	175-225	90	f 7 / s 30	4 / 85							39 / 185	20 / 350				
	225-275	65														

The problem below makes reference to this part of Table 11.

6

Use Table 11 and the formula:

$$RPM = \frac{4 \times cutting\ speed}{diameter}$$

What is the RPM for milling a keyway in a 4.0 diameter 1020 steel shaft with a Bhn of 100? The cutter is .375 diameter high speed steel end mill.

In this question, the important information is the type of steel, the cutter material and diameter of the end mill. The diameter of the shaft is of no concern. Insert known values into the RPM equation:

$$RPM = \frac{4 \times 110}{.375}$$

$$RPM = 1173.3$$

Analyze, Evaluate, & Implement

What is the feed rate in inches per minute (ipm) for this operation if the end mill has four flutes?

Use the formula:

ipm = RPM × feed per tooth × number of teeth on the cutter

Machining Operations

Fill in the known values.

$$ipm = 1173.3 \times .004 \times 4$$

$$ipm = 18.7$$

Table 11 shows the chip load (or feed per tooth) as "4" as Avg (average) and "7" as Opt (optimum). These values are given in thousandths to save room on the table. The average feed per tooth of .004 was chosen with the understanding that the feed can be increased if cutting conditions permit it. The machinist monitors the color of the metal shavings (chips), how the process sounds, and personal experience to make adjustments.

Safety First

• **Clamp the workpiece securely when performing milling operations. Check the tightness of the vice mounting fasteners.**

6

Shop Recommended pages 1060–1062, Table 17

Free-machining steel has a cutting speed of about 100.

Table 17. Feeds and Speeds for Drilling, Reaming, and Threading Plain Carbon and Alloy Steels

Material	Brinell Hardness	Drilling HSS Speed (fpm)	Reaming HSS Speed (fpm)	f/s	Drilling HSS Opt.	Drilling HSS Avg.	Indexable Insert Coated Carbide Opt.	Indexable Insert Coated Carbide Avg.	Reaming HSS Opt.	Reaming HSS Avg.	Threading HSS Opt.	Threading HSS Avg.
Free-machining plain carbon steels (Resulfurized): 1212, 1213, 1215	100-150	120	80	f	21	11	8	4	36	18	83	20
				s	55	125	310	620	140	185	140	185
	150-200	125	80									
(Resulfurized): 1108, 1109, 1115, 1117, 1118, 1120, 1126, 1211	100-150	110	75	f	16	8	8	4	27	14	83	20
	150-200	120	80	s	50	95	370	740	105	115	90	115
	175-225	100	65	f			8	4				
				s			365	735				
(Resulfurized): 1132, 1137, 1139, 1140, 1144, 1146, 1151	275-325	70	45									
	325-375	45	30									
	375-425	35	20									
(Leaded): 11L17, 11L18, 12L13, 12L14	100-150	130	85									
	150-200	120	80									
	200-250	90	60	f			8	4				
				s			365	735				
Plain carbon steels: 1006, 1008, 1009, 1010, 1012, 1015, 1016, 1017, 1018, 1019, 1020, 1021, 1022, 1023, 1024, 1025, 1026, 1513, 1514	100-125	100	65	f	21	11	8	4	36	18	83	20
				s	55	125	310	620	140	185	140	185
	125-175	90	60									
	175-225	70	45	f			8	4				
	225-275	60	40	s			365	735				
Plain carbon steels: 1027, 1030, 1033, 1035, 1036, 1037, 1038, 1039, 1040, 1041, 1042, 1043, 1045, 1046, 1048, 1049, 1050, 1052, 1524, 1526, 1527, 1541	125-175	90	60									
	175-225	75	50									
	225-275	60	40	f			8	4				
	275-325	50	30	s			365	735				
	325-375	35	20									
	375-425	25	15									

f = feed (0.001 in./rev), s = speed (ft/min) *Metric Units:* f × 25.4 = mm/rev, s × 0.3048 = m/min

The last two numbers in this designation refers to how much carbon is in the steel. As the carbon content increases, machineability decreases.

This is the surface feet per minute (sfm) that allows the most efficient machining.

SECTION 6

Machining Stainless Steel

There are dozens of different types of stainless steel, but they all share chromium as an alloying element. Chromium is what makes stainless steels corrosion resistant. Chromium can make a material difficult to machine. The addition of nickel as an alloying element compounds the problem by making the material gummy. Machining characteristics can be improved by the addition of sulfur as an alloying element. The best way to machine stainless steel is to follow proven techniques which begin by identifying the type of stainless steel and its hardness.

Based on the material's characteristics, stainless steel can be placed into one of three categories, as seen in Table 6.1:

Table 6.1 Stainless Steel-Characteristics and Machinability

	Group 1	Group 2	Group 3
Type	Martensitic	Ferritic	Austenitic
Type no.	403, 410, 414, 416, 420, 440A, 440B, 440C, 501	405, 406, 409, 429, 430, 434, 436, 442, 446, 502	201, 202, 301, 302, 303, 304, 304L, 305, 308, 316, 321, 347, 348
Magnetic?	Yes	Yes	No
Machineability	Sulfur types machine without difficulty–similar to medium alloy steel. Can work-harden causing rapid tool wear and poor finish.	Similar to Group 1	Difficult to machine due to nickel content. Workhardens easily. Build-up on cutter tip causes poor finish. Cutting tools wear rapidly.
Suggestions for success	Maintain ridged setup. Use flood coolant. Climb mill if machine permits.	Similar to Group 1	Do not take light cuts (.06 minimum). Climb mill if machine permits. Maintain chipload of at least .004 for milling; .010 for turning. Do not release cutting feed while machining. Use flood coolant. Clamp workpiece securely.

Navigation Hint: Table 6.1 was created for this guide; *it is not found in Machinery's Handbook.*

Machineability of stainless steel is greatly influenced by its hardness. In Table 13 below, the cutting speed of stainless steel (expressed as fpm or "feet per minute") varies greatly depending on the hardness of the material.

Shop Recommended Table 13, page 1049

Table 13. Cutting Feeds and Speeds for Milling Stainless Steels

Material	Brinell Hardness	HSS Speed (fpm)		End Milling									Face Milling		Slit Milling			
				HSS		Uncoated Carbide		Coated Carbide			Coated Carbide			Uncoated Carbide		Coated Carbide		
				Opt.	Avg.	Opt.	Avg.	Opt.	Avg.		Opt.	Avg.		Opt.	Avg.	Opt.	Avg.	
				f = feed (0.001 in./tooth), s = speed (ft/min) *Metric Units:* f × 25.4 = mm/rev, s × 0.3048 = m/min														
Free-machining stainless steels (Ferritic): 430F, 430FSe	135-185	110	f	7	4	7	4	7	4		39	20		39	20	39	20	
			s	30	80	305	780	420	1240		210	385		120	345	155	475	
(Austenitic): 203EZ, 303, 303Se, 303MA, 303Pb, 303Cu, 303 Plus X	135-185	100																
	225-275	80	f	7	4	7	4							39	20			
(Martensitic): 416, 416Se, 416 Plus X, 420F, 420FSe, 440F, 440FSe	135-185	110	s	20	55	210	585							75	240			
	185-240	100																
	275-325	60																
	375-425	30																
Stainless steels (Ferritic): 405, 409, 429, 430, 434, 436, 442, 446, 502	135-185	90	f	7	4	7	4	7	4		39	20		39	20	39	20	
			s	30	80	305	780	420	1240		210	385		120	345	155	475	
(Austenitic): 201, 202, 301, 302, 304, 304L, 305, 308, 321, 347, 348	135-185	75																
	225-275	65																
(Austenitic): 302B, 309, 309S, 310, 310S, 314, 316, 316L, 317, 330	135-185	70	f	7	4	7	4							39	20			
			s	20	55	210	585							75	240			
(Martensitic): 403, 410, 420, 501	135-175	95																
	175-225	85																
	275-325	55																
	375-425	35																

6

Brinell Hardness refers to the material's resistance to penetration. The higher the Brinell Hardness, the lower the machineability. Cutting speeds of 100 are comparable to regular machine steel. Cutting speeds of 35, as seen above, require much lower spindle RPMs. Determining the material's hardness is a very important step in setting up a machine tool to cut stainless steel.

Determine the Hardness of a Material

Index navigation path and key words:

— hardness/testing/Brinell/page 505

The Brinell hardness scale is one standard used to report the hardness of a metal. A steel ball penetrator is pressed into the metal with a predetermined load. The diameter of the resulting impression is measured and translated to the values seen in Table 13, "Brinell Hardness." Softer materials produce larger diameters. Using Brinell Hardness to determine the cutting speed of a metal is very reliable, but sometimes hardness testing is not possible.

Making It Simple

One of the oldest and most popular methods for determining a metal's hardness is the "file test." The way a file "bites" or drags when filing steel can give an approximate hardness value. Soft metal that files easily has a Bhn of about 100. Hardened steel will cause a file to glide across it like glass because the file and the steel are about the same hardness. Steel of a known hardness can be compared to unknown samples to determine an approximate hardness value. Hardness testing files of different hardness's are available for this very purpose.

File testing takes practice and experience; the result does not give an absolute value.

Determine the RPM for milling 303 stainless steel with a .75 diameter end mill and a cutting speed of 100.

For milling, the RPM formula looks like this:

$$RPM = \frac{4 \times cutting\ speed}{diameter\ of\ cutter}$$

Fill in the known values:

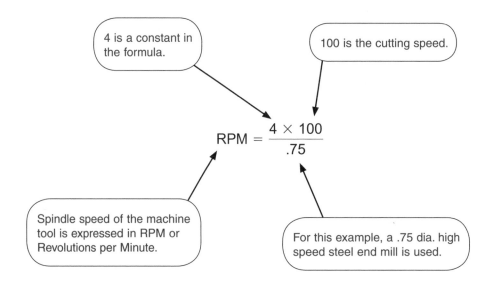

Solve the equation:

$$RPM = 533.3$$

More suggestions for machining stainless steels:

- Use flood coolant.
- Do not allow the cutter to dwell during cutting.
- If cutter breakdown is excessive, try *increasing* the feed rate.
- "Climb Mill" if the machine permits this type of operation.
- Clamp the workpiece securely!

Unit F: Cutting Fluids pages 1182–1191

Index navigation path and key words:
— cutting fluids/for different materials/page 1184

Cutting fluids can be used to improve the machining process by cooling and providing lubrication. Other benefits include extended cutting tool life, better surface finishes, the ability to take deeper cuts at higher speeds, and flushing away chips. Many machine tools have a fluid pump, a reservoir, and a filtering system that work together to circulate the coolant. *Machinery's Handbook* identifies four major divisions of cutting fluids:

- Cutting oils
- Water-miscible fluids
- Gases
- Paste and solid lubricants

Cutting Oils
Cutting oils have the appearance of motor oil and have chemicals added to improve lubricity. Cutting oils have superior lubricating properties, but do not cool as well as water-based cutting fluids. Uses include gear cutting, tapping, and thread grinding.

Water-Miscible Fluids
Emulsions or soluble oils combine the lubricating and rust prevention properties of oil with the cooling ability of water. *Chemical fluids* are characterized by their high cooling, high rust prevention properties. Their lubricating properties are less than *Emulsions or soluble oils. Semichemical fluids* are combinations of chemical fluids and *emulsions.* Uses for water-miscible fluids include grinding operations, milling, and turning operations and drilling.

Gases
Gases are introduced at the cutter-workpiece interface under pressure. Gases include:

- Air
- Freon
- Nitrogen

A common use for gases is simply a well-directed air blast pointed at the machining operation. Other systems introduce Freon though the tools directly to the tool-workpiece interface for maximum effect.

Paste and Solid Lubricants

Paste and solid lubricants such as waxes, soaps, graphite, and molybdenum disulfide are applied directly to the tool or workpiece, or impregnated in the cutting tool. There use is limited to special applications.

Safety First

Chemicals used in the shop are required to have a Material Safety Data Sheet (MSDS)

A Material Safety Data Sheet (MSDS) is designed to provide both workers and emergency personnel with the proper procedures for handling or working with a particular substance. MSDS's include information such as physical data (melting point, boiling point, flash point etc.), toxicity, health effects, first aid, reactivity, storage, disposal, protective equipment, and spill/leak procedures.

Analyze, Evaluate, & Implement

Find the MSDS document for a chemical used in the shop.
Explain the precautions listed in the MSDS sheet

Unit I: Grinding and Other Abrasive Processes
pages 1216–1278

 **Index navigation path and key words:**

— grinding/surface/pages 1261–1267

Grinding is a metal removal process that uses a rotating abrasive wheel. Close tolerances, the ability to machine hard materials, and superior surface finishes are characteristics of grinding. The abrasive grains of the grinding wheel removes tiny chips of material similar to the milling process. Common types of grinding machines are:

- Surface Grinder
- Cylindrical Grinder
- Internal Grinder
- Tool Grinder
- Gear Grinder

Surface Grinder

The horizontal spindle surface grinder is a very common machine tool. The workpiece is often held by a magnetic chuck which is a flat table that reciprocates back and forth under the rotating grinding wheel. The table can also be fed in and out (toward and away from the operator). The spindle carrying the grinding wheel can be moved up and down in precise increments.

Surface Grinders are identified by their table size. A small machine may be a 6–12 meaning it has a table that is 6″ × 12″. Larger machines can have tables several feet long.

Cylindrical Grinder

Cylindrical grinders are used to grind the outside diameters and faces of cylindrical-shaped workpieces. On universal cylindrical grinders, the part is held between 60-degree machined centers or held by a chuck. By rotating the table of the machine slightly, tapers can be ground. Some machines have an attachment to grind internal diameters. Centerless grinders are production machines used to grind the outside diameter of multiple pieces to the same diameter such as dowels.

Internal Grinder, Tool Grinder, Gear Grinder

There are other grinders that belong to a collection of machines used for special purposes. These machines are set up to perform a specific operation very efficiently. *Internal grinders* are used to grind precise internal diameters for bearing races, hydraulic pumps, and cylinders to name a few. *Gear grinders* have the complex shape of the gear tooth dressed on the grinding wheel. The machine accurately indexes the gear blank to expose each tooth for finishing. *Tool grinders* are multi-axis machines that produce cutting tools from solid blanks of high speed steel or carbide. Tool grinding shops supply the tool and die industry with custom made cutters having special profiles for unique applications.

6

Abrasives

 Index navigation path and key words:

— abrasive/grinding/page 1216

Grinding wheels are made from abrasive grains. Common abrasive grains are:

- Aluminum Oxide
- Silicon Carbide
- Cubic Boron Nitride (CBN)
- Diamond

Aluminum Oxide

Aluminum oxide grinding wheels are the most common grinding wheel in manufacturing. They are made from grains of abrasive grit held together by a bond that leaves spaces between the individual grains. During the grinding process grains break off exposing new sharp particles. Aluminum oxide wheels are used on surface grinders, cylindrical grinders and pedestal grinders. These wheels are for general purpose grinding and work well on various types of steel.

Silicon Carbide

Silicon carbide wheels are similar to aluminum oxide wheels in that they are made from abrasive grains and held together with a bonding agent, but they are harder than aluminum oxide wheels. They work well on cast iron and non-ferrous materials such as aluminum and bronze.

Cubic Boron Nitride (CBN or Borazon)

In the 1950s General Electric developed a super abrasive called *Borazon*. In recent history, this material was improved to make it more versatile. Unlike "stone" wheels, the abrasive is applied to a thin layer on the outside diameter of the wheel. It works well on materials that have a large amount of alloying elements (super alloys) and hard-to-machine nickel alloys.

Diamond

Diamond grinding wheels can be made from natural diamonds, but most are synthetic. Like *CBN* wheels, the abrasive is applied to a thin layer on the outside of the wheel. *Diamond* grinding wheels are used to grind very hard materials such as carbide, glass, ceramics, and cement products.

Truing and Dressing

Truing is the process of making the grinding wheel concentric to the machine spindle. Aluminum oxide and silicon carbide wheels or "stone" wheels are *dressed* periodically with a diamond to remove pieces of workpiece material that become trapped in the grinding wheel and to expose fresh, sharp abrasive particles. In this case, dressing also trues the wheel. The diamond is held in a steel block that is placed on the magnetic table of the grinder to the left of the centerline of the wheel. Because the grinding wheels rotate clockwise, the diamond is not in danger of becoming trapped by the wheel. Four or five passes of .001 is usually enough to freshen the wheel.

Diamond and *CBN* wheels are dressed with a hand-held stick that cleans the wheel. In this case, dressing does not accomplish truing.

Machining Operations

Grinding Wheel Shapes

 Index navigation path and key words:

— grinding/wheels/pages 1216–1239, 1246–1250

Grinding wheels are available in 28 different configurations for different types of applications. These are the types that are used most often in the typical machine shop:

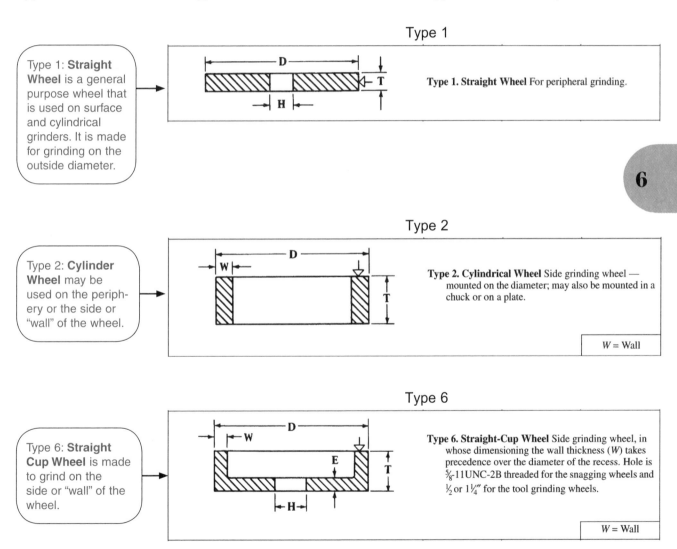

Type 1: **Straight Wheel** is a general purpose wheel that is used on surface and cylindrical grinders. It is made for grinding on the outside diameter.

Type 1

Type 1. Straight Wheel For peripheral grinding.

Type 2: **Cylinder Wheel** may be used on the periphery or the side or "wall" of the wheel.

Type 2

Type 2. Cylindrical Wheel Side grinding wheel — mounted on the diameter; may also be mounted in a chuck or on a plate.

W = Wall

Type 6: **Straight Cup Wheel** is made to grind on the side or "wall" of the wheel.

Type 6

Type 6. Straight-Cup Wheel Side grinding wheel, in whose dimensioning the wall thickness (W) takes precedence over the diameter of the recess. Hole is $\frac{5}{8}$-11UNC-2B threaded for the snagging wheels and $\frac{1}{2}$ or $1\frac{1}{4}''$ for the tool grinding wheels.

W = Wall

6

145

Type 11

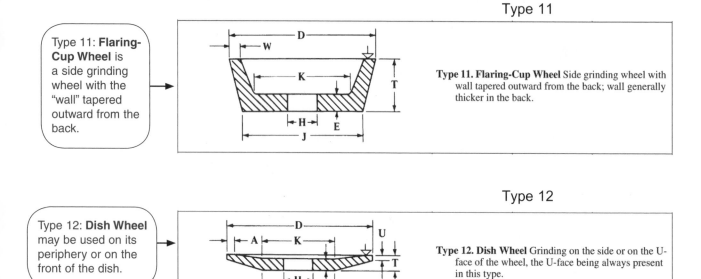

Type 11: **Flaring-Cup Wheel** is a side grinding wheel with the "wall" tapered outward from the back.

Type 11. Flaring-Cup Wheel Side grinding wheel with wall tapered outward from the back; wall generally thicker in the back.

Type 12

Type 12: **Dish Wheel** may be used on its periphery or on the front of the dish.

Type 12. Dish Wheel Grinding on the side or on the U-face of the wheel, the U-face being always present in this type.

6

Safety First

- Handle grinding wheels carefully, do not drop.
- Cracked grinding wheels are dangerous and should be discarded.
- "Ring test" a grinding wheel before mounting by supporting the wheel on a shaft and tapping the wheel with a plastic screw driver handle. A good wheel sounds like a bell; a bad wheel sounds like a thud.
- Always use the mounting blotters (gaskets) between the wheel and the hub.
- Allow a new grinding wheel to run for a minute before using to make sure there are no cracks in the wheel that may cause it to explode.
- Never use a machine with machine guards removed.

Grinding Wheel Markings

 Index navigation path and key words:
— grinding/wheels/markings/standard/pages 1218–1219

Grinding wheels are identified by a standardized five–position marking system that gives important information about the wheel. Sometimes there are numbers before position 1 or after position 6. These are manufacturer's codes and can be disregarded by the machinist.

Prefix	1 Abrasive Type	2 Grain Size	3 Grade	4 Structure	5 Bond Type	6 Manufacturer's Record
51 –	**A** –	**36** –	**L** –	5 –	**V** –	23

Position 1   Abrasive type
- (A)   Aluminum Oxide
- (C)   Silicon Carbide
- (D)   Diamond
- (SG)   Ceramic Aluminum Oxide
- (MD)   Manufactured Diamond
- (B)   Cubic Boron Nitride (CBN)

Position 2   Grain Size
For wheels used in machine shops, this position is 36–220. 36 is a coarse wheel, 220 is fine. Coarse wheels are for softer material; fine wheels are for hard materials such as hardened steel.

Position 3   Grade of Hardness
Indicated by letters of the alphabet from A–Z. During grinding, abrasive grains are pulled from soft wheels more readily than hard wheels.

Position 4   Structure
The structure is indicated by numbers 1–16. Higher numbers indicate wider spaces between grains. High structure numbers do not load up with material as easily. Low structure numbers provide greater detail and sharper corners in the workpiece.

Position 5   Bond
The type of glue that holds the abrasive grains together.
- (V)   Vitrified
- (S)   Silicate
- (E)   Shellac
- (R)   Rubber
- (B)   Resinoid
- (M)   Metal

6

Problems and Solutions in Surface Grinding

 Index navigation path and key words:

— grinding/surface grinding troubles/page 1266

There are probably more variables in precision grinding than in any other machining operation. Consequently, there are many things that go wrong. Table 4 on page 1198 is an effective tool for diagnosing the most frequent challenges you may face while grinding.

Shop Recommended page 1267, Table 4

> The left hand column gives solutions to the problems shown on the top of the table.

> The top of the table gives the most common problems encountered when grinding.

Table 4. Common Faults and Possible Causes in Surface Grinding

CAUSES	FAULTS	Work not flat	Work not parallel	Poor size holding	Burnishing of work	Burning or checking	Feed lines	Chatter marks	Scratches on surface	Poor finish	Wheel loading	Wheel glazing	Rapid wheel wear	Not firmly seated	Work sliding on chuck
		WORK DIMENSION			METALLURGICAL DEFECTS		SURFACE QUALITY				WHEEL CONDITION			WORK RETAINMENT	
WORK CONDITION	Heat treat stresses	●													
	Work too thin		●												
	Work warped	●												●	
	Abrupt section changes	●	●												
GRINDING WHEEL	Grit too fine				●	●					●	●			
	Grit too coarse									●					
	Grade too hard	●				●		●			●	●			
	Grade too soft			●		●		●	●				●		
	Wheel not balanced							●							
	Dense structure										●	●			
TOOLING AND COOLANT	Improper coolant										●				
	Insufficient coolant	●			●	●						●			
	Dirty coolant								●						
	Diamond loose or chipped	●	●						●						
	Diamond dull			●						●					
	No or poor magnetic force			●										●	●
	Chuck surface worn or burred	●	●						●					●	●
MACHINE AND SETUP	Chuck not aligned	●	●												
	Vibrations in machine							●							
	Plane of movement out of parallel	●	●												
OPERATIONAL CONDITIONS	Too low work speed										●				
	Too light feed											●			
	Too heavy cut	●								●					
	Chuck retained swarf	●	●		●										●
	Chuck loading improper	●	●												●
	Insufficient blocking of parts									●					
	Wheel runs off the work		●	●									●		
	Wheel dressing too fine	●													
	Wheel edge not chamfered						●								
	Loose dirt under guard								●						

> Example: The grinding wheel is leaving "burn marks" in the workpiece. Burn marks look like thin brown stripes. These are known as "Metallurgical Defects" because the metal has actually changed in the area of the burn.

> Possible causes are indicated by a dot.
> In the **Grinding Wheel** section:
> • Grit too fine
> • Grade too hard
> • Tooling and coolant:
> • Insufficient coolant.
> • Operational Conditions:
> • Wheel runs off work

Safety First

- Grinding wheels on surface grinders turn clockwise. Stand to the right of the machine spindle when operating the machine to avoid being in the path of flying debris.
- Surface grinder wheels are conditioned by "dressing" the wheel with an industrial diamond to provide sharp cutting particles. Position the diamond on the machine table to the left of the spindle centerline so the rotation of the wheel does not pull the diamond into a tight area.
- Never touch a rotating grinding wheel.

Unit J: CNC Numerical Control Programming
pages 1279–1318

6

 Index navigation path and key words:

— CNC/G codes/page 1284

Computer Numerical Control Programming—better known as *CNC programming*—is a vital part of technology today. Combinations of letters of the alphabet and numbers are used in a specific coded sequence to control the machine tool. This language is known as *G code*. Programs can be used repeatedly to produce identical results.

CNC Coordinate Geometry

A CNC machine interprets data and translates it into machine movements in multiple axes. On a vertical milling machine, the *X* axis is to right and left of the operator, the *Y* axis is toward and away from the operator, and the *Z* axis is the spindle (or up and down). The *Cartesian coordinate system* is used to define locations from the origin, known as *X*, *Y*, and *Z* positions.

SECTION 6

Shop Recommended page 1280, Figures 1 and 2

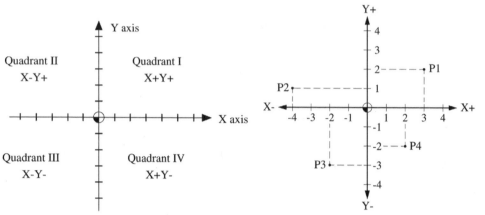

Fig. 1. Rectangular Coordinate System.   Fig. 2. Absolute Coordinates.

In Figure 1, the origin is the intersection of the *X* and *Y* axes. Locations to right of and above the origin are positive numbers known as *X+ Y+*; locations to the left of and above the origin are *X− Y+*; to the left of and below, *X− Y−*; to the right and below, *X+ Y−*. *X*, *Y*, and *Z* locations are a given distance from the origin shown as the intersection of the *X* and *Y* axes. In Figure 2, P1 is defined as *X*3.0, *Y*2.0; P2 is *X−*4.0, *Y*1.0; P3 is *X−*2.0, *Y−*3.0; P4 is *X*2.0, *Y−*2.0.

Shop Recommended page 1280, Figure 3

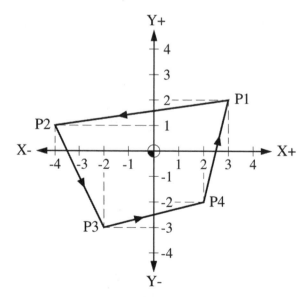

                  Fig. 3. Incremental Motions.

Machining Operations

Once point locations are determined, the programmer can use them as points representing the center of holes, machined edges, or a contour. A *toolpath* can be created by connecting points. Imagine that points P1, P2, P3, and P4 represent the path of a milling cutter. Table 6.2 is a coordinate list showing *X* and *Y* positions relative to the origin:

Table 6.2 Coordinate List

	X	Y
P1	3.0	2.0
P2	−4.0	1.0
P3	−2.0	−3.0
P4	2.0	−2.0

G-codes are the language of CNC machines. The G-codes below are standard for most–but not all–machine controls. Some G-codes are *modal commands*. *Modal* commands stay in effect until changed or canceled. Conflicting codes cannot be used in the same line of information. Lines are separated by an EOB End of Block sign which is a (;).

6

Analyze, Evaluate, & Implement

Plot your initials on a coordinate list

Materials needed:

- Graph paper with four squares per inch
- Mechanical pencil with .5mm HB lead
- Straight edge or 6″ steel scale

1. With the graph paper in the "landscape" position, draw a rectangle 4″ × 6″.

2. Identify the upper left corner as "0–0".

3. Identify the center of the rectangle as "X 3.0; Y2.0".

4. Center the three letters of your initials in the rectangle. Letters are to be 1 inch wide by 2 inches tall with 1/2″ in between letters. Use straight lines; do not use arcs.

5. Make a coordinate list similar to Table 6.3 showing the points identifying your initials.

X and Y "0"

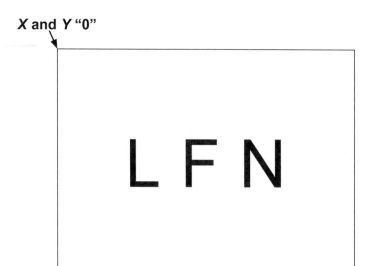

Table 6.3 Coordinate List for Your Initials

Coordinate List

X	Y

Index navigation path and key words:
— G-address, CNC programming/page 1284

Shop Recommended page 1286, Table 2

1286   CNC G-CODES

Table 2. Typical Milling G-codes

G- code	Description	G- code	Description
G00	Rapid positioning	G52	Local coordinate system setting
G01	Linear interpolation	G53	Machine coordinate system
G02	Circular interpolation clockwise	G54	Work coordinate offset 1
G03	Circular interpolation counterclockwise	G55	Work coordinate offset 2
G04	Dwell (as a separate block)	G56	Work coordinate offset 3
G09	Exact stop check - one block only	G57	Work coordinate offset 4
G10	Programmable data input (Data setting)	G58	Work coordinate offset 5
G11	Data setting mode - cancel	G59	Work coordinate offset 6
G15	Polar coordinate command - cancel	G60	Single direction positioning
G16	Polar coordinate command	G61	Exact stop mode
G17	XY plane designation	G62	Automatic corner override mode
G18	ZX plane designation	G63	Tapping mode
G19	YZ plane designation	G64	Cutting mode
G20	U.S. customary units of input (G70 on some controls)	G65	Custom macro call
G21	Metric units of input (G71 on some controls)	G66	Custom macro modal call
G22	Stored stroke check ON	G67	Custom macro modal call - cancel
G23	Stored stroke check OFF	G68	Coordinate system rotation

152   continued on next page

G25	Spindle speed fluctuation detection ON	G69	Coordinate system rotation - cancel
G26	Spindle speed fluctuation detection OFF	G73	High speed peck drilling cycle (deep hole)
G27	Machine zero position check	G74	Left hand threading cycle
G28	Machine zero return (reference point 1)	G76	Fine boring cycle
G29	Return from machine zero	G80	Fixed cycle - cancel
G30	Machine zero return (reference point 2)	G81	Drilling cycle
G31	Skip function	G82	Spot-drilling cycle
G40	Cutter radius offset - cancel	G83	Peck-drilling cycle (deep hole drilling cycle)
G41	Cutter radius offset - left	G84	Right hand threading cycle
G42	Cutter radius offset - right	G85	Boring cycle
G43	Tool length offset - positive	G86	Boring cycle
G44	Tool length offset - negative	G87	Back boring cycle
G45	Position compensation - single increase (obsolete)	G88	Boring cycle
G46	Position compensation - single decrease (obsolete)	G89	Boring cycle
G47	Position compensation - double increase (obsolete)	G90	Absolute dimensioning mode
G48	Position compensation - double decrease (obsolete)	G91	Incremental dimensioning mode
G49	Tool length offset cancel	G92	Tool position register
G50	Scaling function cancel	G98	Return to initial level in a fixed cycle
G51	Scaling function	G99	Return to R level in a fixed cycle

6

M-Codes

Index navigation path and key words:

— M codes, CNC/page 1287

M-codes are *machine codes* that command machine functions such as turning on the spindle or coolant.

Shop Recommended page 1287, Table 3

Table 3. M-codes

M code	Description	M code	Description
M00	Program stop	M09	Coolant OFF
M01	Optional program stop	M19	Spindle orientation
M03	Spindle rotation normal (clockwise)	M30	Program end
M04	Spindle rotation reverse (counterclockwise)	M60	Automatic Pallet Change (APC)
M05	Spindle stop	M98	Subprogram call
M06	Automatic Tool Change (ATC)	M99	Subprogram end
M08	Coolant ON		

Shop Recommended page 1280, Figure 7

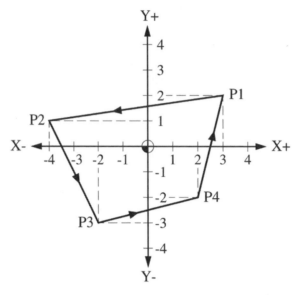

Figure 7

Figure 7 will be used to write a simple program using g-codes. Refer to page 1286, Table 2, *"CNC G-codes"* and page 1287, Table 3, *"M Codes"* while studying this program. If the bold lines represent the cutter path, the program for the shape shown would look like this:

G-Code	Meaning
T01 M06	(tool number 1; change tool)
G00 G54 G90 X3.0 Y2.0 S1000 M03	(rapid to work coordinate offset 1, X3.0 Y2.0; use absolute coordinates; spindle speed 1000 rpm clockwise)[P1]
G43 Z.1 H01	(tool length offset position 1; rapid tool to .1 above the part)
G01 Z-.03 F10.0	(feed down into the part .03 at a feed of 10.0 Inches per minute)
X-4.0 Y1.0	(feed to this position) [P2]
X-2.0 Y-3.0	(feed to this position) [P3]
X2.0 Y-2.0	(feed to this position) [P4]
X3.0 Y2.0	(feed to this position) [P1]
Z.10	(feed up to .1 above the part)
G28	(go to machine "0")
M30	(program end)

Machining Operations

Shop Recommended page 1322, Table1

Table 1. Degrees of Accuracy Expected with NC Machine Tools

Type of NC Machine	Accuracy	
	inches	mm
Large boring machines or boring mills	0.0010-0.0020	0.025-0.050
Small milling machines	0.0006-0.0010	0.015-0.025
Large machining centers	0.0005-0.0008	0.012-0.020
Small and medium-sized machining centers	0.0003-0.0006	0.008-0.015
Lathes, slant bed, small and medium sizes	0.0002-0.0005	0.006-0.012
Lathes, small precision	0.0002-0.0003	0.004-0.008
Horizontal jigmill	0.0002-0.0004	0.004-0.010
Vertical jig boring machines	0.0001-0.0002	0.002-0.005
Vertical jig grinding machines	0.0001-0.0002	0.002-0.005
Cylindrical grinding machines, small to medium sizes	0.00004-0.0003	0.001-0.007
Diamond turning lathes	0.00002-0.0001	0.0005-0.003

6

Unit C: Estimating Speeds and Machining Power pages 1081–1090, 1052–1061

Engineers will find this unit helpful for conducting time studies on machining operations that involve multiple parts and repeat orders. When purchasing a machine tool, one of the most important considerations is the horsepower at the spindle. This unit provides necessary formulas for determining machine power.

 ## Index navigation paths and key words:

Unit D: Micromachining pages 1092–1131

The recent advancements in the area of product miniaturization have inspired a new Unit in MH 29 "Machining Operations." Micromachining does not have a standardized definition at this time, but it is typically used to produce components with dimensions of less than 1 mm (.04) or when the depth of cut is comparable to tool sharpness or tool grain size. There are machine tools commercially available in the area of micromanufacturing, but a large part this new technology is done on the theoretical level in universities and research facilities. It is not possible to simply extend macroscale machining practices to the microscale level.

SECTION 6

Characteristics of Micromachining:

- Workpiece or spindle speeds of 25,000 RPM or higher
- Spindle runout controlled to the submicron level (runout, tool concentricity, and positioning accuracy near 1/100 of tool diameter)
- Strong mechanical and thermal structure that is unaffected by vibration or thermal drift
- Highly accurate and controllable feeding mechanisms

 Index navigation paths and key words:

Unit E: Machine Econometrics pages 1132–1168

Econometrics is the science of using mathematics and statistical data to predict an outcome such as tool life. This information is then used to calculate costs.

 Index navigation paths and key words:

Unit F Screw Machines, Band Saws pages 1170–1191

A screw machine is a production turning machine that produces multiple parts, sometimes sold by weight because of the sheer volume of parts. Machining operations common to a screw machine are threading, boring, form cutting, drilling, reaming, and parting off.

Safety First

**A hazard common to the turning machine is a "wrap point,"
the area between the rotating chuck and the machine bed.**

Index navigation paths and key words:

Unit G: Machining Nonferrous Metals and Non-Metallic Materials pages 1192–1196

Metal is a material that reflects light and conducts electricity. Nonferrous metals are metals that do not contain iron. Examples of nonferrous metals are: magnesium, copper, brass, lead, and zinc. These metals have special properties that affect the machining process.

Index navigation paths and key words:

6

More information is found in Section 6, "Machine Operations," under:

Safety First

Magnesium is a non-ferrous element that burns furiously! Never use water to put out a magnesium fire. When machining magnesium, check the Material Safety Data Sheet (MSDS) for precautions.

Unit H: Grinding Feeds and Speeds pages 1197–1215

The speeds of most grinding machines in the typical shop are not adjustable, so calculating speeds for grinding is not a common practice. On a surface grinder, the controllable variables are depth of cut and table speed. The design of grinding machines is such that they will accept only wheel diameters of a certain size.

SECTION 6

Index navigation paths and key words:

— grinding/minimum cost conditions/page 1200
— grinding/ECT/grinding/page 1198
— grinding/wheel life/wheel life/life vs. cost/page 1146

For more information on Grinding Feeds and Speeds, see

— feeds and speeds/grinding/page 1215

ASSIGNMENT

List the key words and give a definition of each:

High Speed Steel	Thermal Shock	Cratering
RPM	Plain Carbon Steel	Surface Feet per Minute
Tool Steel	Ferrous and Nonferrous	
Cutting Speed	Alloy Steel	

APPLY IT!

For questions 1–3, refer to the tables indicated to confirm the information shown in the question.

1. Refer to page 1027, Table 1. What is the RPM and feed rate for a turning machine if the material is 4.50 diameter 1060 plain carbon steel using uncoated carbide inserts? The cutting speed is 300. Use the formula 4 × the cutting speed divided by the diameter.

2. Refer to page 1031, Table 3. What is the RPM and feed rate for a lathe if the material is .375 diameter; 304 stainless steel? The cutter is high speed steel. The Brinell Hardness is 140. The cutting speed is 75. Use the formula 4 × the cutting speed divided by the diameter.

3. What is the RPM for drilling 1020 plain carbon steel with a letter "K" HSS drill with a cutting speed of 60? Use the Speeds and Feeds chart beginning on page 1016 to determine the RPM. What would the RPM be for a 15/16 drill with a cutting speed of 90?

4. How is feed expressed for a milling machine? A turning machine?

5. What is the result of using a grinding wheel that is too hard?

6. What is the range of grit size used in the machine shop?

Machining Operations

7. Interpret the symbols on this grinding wheel:

C   46   J 6   V

8. What are the possible causes of scratches in the workpiece when using a surface grinder?

   Questions 9 and 10 refer to the CNC milling machine.

9. Give three examples of M codes.

10. What G code is used to feed from one point to another point?

11. A milling machine is cutting mild steel with a four flute 3/4″ diameter end mill operating at 480 RPM and a feed rate of 7.68 ipm. What is the thickness of the chips being produced in this operation?

12. If the cutter in question 11 is changed to a two flute end mill, what is the chip thickness?

13. How much is the diameter of a workpiece reduced if the cutting tool on the lathe is fed in .015?

14. What is the formula for determining inches per minute (ipm) for a drill given the inches per revolution (ipr)?

   Refer to Table 1 and Table 2 for questions 15–30.

Table 1

Feed in inches per revolution for drills	
0-1/8	.001
1/8-1/4	.003
1/4-3/8	.004
3/8-1/2	.006
1/2-1.0	.010

Table 2

Workpiece Material	Cutting Tool Material	
	HSS	**Carbide**
Aluminum	300	1500
Mild Steel (1018)	90	300
Alloy Steel (4140; 4340)	60	240
High Carbon Steel (1040; 1060)	50	150
Stainless Steel	50	150

SECTION 6

15. What is the feed in ipr for a for a "C" drill turning at 1322 RPM?

16. What is feed in ipr for a "U" drill turning at 543 RPM?

17. Feeds expressed in "inches per minute" are commonly found on what type of machine tool?

18. What is the formula for determining the feed rate for a milling machine?

19. What is the RPM for a 3.0 dia. carbide face mill machining aluminum?

20. What is the RPM for 1/4″ four flute HSS end mill cutting alloy steel?

21. The RPM on a milling machine is set at 100. The cutter is a 6″ diameter face mill with 8 carbide inserts. What is the surface feet per minute (SFM) for this operation?

22. What is the formula for determining the feed rate in ipm for a milling machine?

23. What is the feed for question #21 if the chip load is .006 per tooth?

24. What is the feed rate for question # 21 if the cutter is changed to a 3.0 diameter carbide face mill with 6 inserts?

25. A 1/4″ keyway is machined in a 2.0 diameter stainless steel shaft with a 2 flute .220 diameter cutter. The chipload or "feed per tooth" is .004. What is the RPM? What is the feed in ipm?

26. What is the feed rate in question #25 if the cutter is changed to a .200 diameter 4 flute end mill?

27. What is the RPM for a 6.0 diameter $\times$ .187 thick HSS slitting saw with 24 teeth? The material is 1018.

28. What is ipm for question #27 if the chip load is .006?

29. What is the thickness of the chips in question #28?

30. What is the RPM for a 1.0 diameter; 4 flute carbide end mill machining aluminum with 1/2″ depth of cut?

6

Section 7

MANUFACTURING PROCESSES

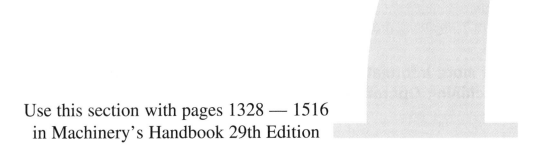

Use this section with pages 1328 — 1516
in Machinery's Handbook 29th Edition

Navigation Overview

All Units in this Section covered with:
Navigation Assistant

The Navigation Assistant helps find information in the MH29 Primary Index. The Primary Index is located in the back of the book on pages 2701–2788 and is set up alphabetically by subject. Watch for the magnifying glass throughout the section for navigation hints.

May I help you?

7

Introduction

The choice of a manufacturing process may mean the difference between making money and losing money. Manufacturing processes selected should be the fastest, safest, and most economical method available, not necessarily the most precise. Here are some examples:

- Wire electrical discharge machining (EDM) gives very accurate results. However, this method is time consuming and expensive compared to others and is not normally used for parts with loose tolerances.
- Stamping dies are a big investment in material, machine time, and labor. These tools are intended for large volumes of parts, sometimes in the millions. For small quantities or for prototypes, laser manufacturing is often used.

When is it practical to invest in die casting? Investment casting? Powder metallurgy? Material in this section of MH will help you to make an informed decision regarding the right manufacturing process for a given job.

For more Information on machine accuracy, see Section 6; Machining Operations, MH29 page 1322

162

Manufacturing Processes

Unit A: Sheet Metal Working and Presses pages 1331–1392

🔍 **Index navigation paths and key words:**

Unit B: Electrical Discharge Machining pages 1393–1403

🔍 **Index navigation paths and key words:**

Unit C: Iron and Steel Castings pages 1404–1423

🔍 **Index navigation paths and key words:**

Unit D: Soldering and Brazing pages 1424–1431

🔍 **Index navigation paths and key words:**

7

SECTION 7

 **Unit E:** Welding pages 1442–1486

Index navigation paths and key words:

— welding/aluminum/page 1460
— welding/ANSI welding symbols/pages 1476–1486
— welding/definitions and symbols/pages 1476–1484
— welding/electrode/characteristics/page 1451
— welding/GMAW/shielding gases/pages 1435, 1437
— welding/GTAW/aluminum/page 1457
— welding/plastics/page 595

Unit F: Lasers pages 1497–1499

Index navigation paths and key words:

— lasers/cutting metal with/page 1491
— lasers/cutting nonmetals/pages 1493–1494
— lasers/types of industrial/page 1489
— lasers/welding/theory/page 1495

 **Unit G:** Finishing Operations pages 1500–1516

Index navigation paths and key words:

— polishing and buffing/page 1501
— polishing and buffing/operations/page 1504
— etching and etching fluids/pages 1505–1506

Analyze, Evaluate, & Implement

Rank five different metal-cutting operations in order of:

- Accuracy
- Surface Finish
- Machining Time
- Cost

Obtain a number of engineering drawings of individual details. Discuss which manufacturing process is the most efficient. Why?

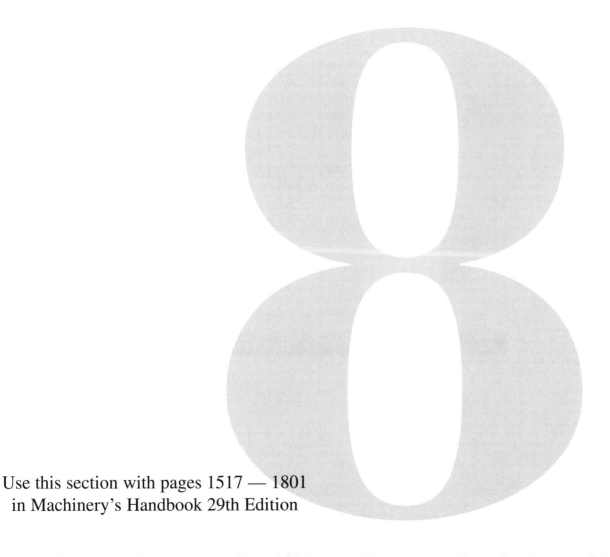

Section **8**

FASTENERS

Use this section with pages 1517 — 1801
in Machinery's Handbook 29th Edition

SECTION 8

Navigation Overview

Units Covered in this Section

Unit G   Cap Screws and Set Screws
Unit K   Pins and Studs
Unit L   Retaining Rings

 Units Covered in this Section with:
Navigation Assistant

The Navigation Assistant helps find information in the MH29 Primary Index. The Primary Index is located in the back of the book on pages 2701–2788 and is set up alphabetically by subject. Watch for the magnifying glass throughout the section for navigation hints.

May I help you?

Unit A   Torque and Tension in Fasteners
Unit B   Inch Threaded Fasteners
Unit C   Metric Threaded Fasteners
Unit D   Helical Coil Screw Threaded Inserts
Unit E   British Fasteners
Unit F   Machine Screws and Nuts
Unit H   Self-Threading Screws
Unit I   T-Slots, Bolts, and Nuts
Unit J   Rivets and Riveted Joints
Unit M   Wing Nuts, Wing Screws, and Thumb Screws
Unit N   Nails, Spikes, and Wood Screws

Key Terms	
Nominal	Bolt
Blind Hole	Screw
Press Fit	Pilot Diameter
Slip Fit	Counterbore
Hexagon Head	Countersink

Learning Objectives

After studying this unit you should be able to:

- Select a fastener for a given application.
- Calculate the counterbore depth for a socket head cap screw.
- Select a set screw for a given application.
- List four applications for dowel holes.
- State seven design guidelines for using dowel pins.
- State two applications for shoulder bolts.

Introduction

Of the multitude of fasteners described in *Machinery's Handbook*, the units covered in this section will focus on the fasteners most commonly used in manufacturing and in the average machine shop. By definition, a *bolt* uses a nut and a *screw* is threaded into a tapped hole. Tools, dies, special machines, fixtures, and gages are assembled with screws because this provides a much stronger assembly that facilitates repair and maintenance. Threaded assemblies using tapped holes are less likely to loosen up and the fit between components is more precise. The thread series of machine screws are usually national coarse (NC) or national fine (NF).

These common fasteners are listed by the way they are expressed in *Machinery's Handbook,* followed by the more common name:

- Hexagon head cap screws (socket head cap screws)
- Hexagon flat countersunk cap screws (flat head cap screws)
- Hexagon button head cap screws (button head cap screws)
- Hexagon head shoulder screws (shoulder bolts or stripper bolts; these are known as "bolts" but are never used with a nut)
- Hexagon socket set screws (set screws)

8

SECTION 8

Socket Head Cap Screws

Index navigation paths and key words:

— cap screws/hexagon socket head/page 1694

Hexagon head cap screws (Socket Head Caps Screws) are abbreviated on engineering drawings as "SHCS". Caps screws pass through a clearance hole and screw into a threaded hole in the mating component. It is a common practice to *counterbore* the clearance hole. *Counterboring* is a machining operation that is used to recess a fastener below the surface of the material. A shallow recess to clean up an area for the head of a fastener is called a *spotface*.

When building a fixture or special tool of your own design, the size and number of fasteners used is a reasonable decision based on the size of the components, the application, and the tool force (if any). Keep in mind that a fastener should have 1 1/2 to 2 times its diameter in *effective* thread depth. Effective thread depth is the actual usable thread not counting the partial lead threads on a tap. It is not uncommon to turn a tap handle four to six revolutions before a full thread is made. Here are examples of recommended thread depths for commonly used fasteners:

- 1/2–13 SHCS   1/2 × 1 1/2 = 3/4″ effective thread depth
- 1/4–20 SHCS   1/4 × 1 1/2 = 3/8″ effective thread depth
- 3/8–16 SHCS   3/8 × 1 1/2 = 9/16″ effective thread depth

Making it Simple

The length of Socket Head Cap Screws is measured from the bottom of the head of the screw to the end of the threads. The height of the head of the screw is always the same as the diameter of the thread. This is helpful for determining the depth of the *counterbore* (see Figure 8.1).

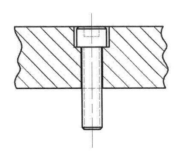

Figure 8.1 Socket Head Cap Screw (SHCS) in a counterbored hole.
The screw head is recessed .03 below the surface of the plate.

Fasteners

Fastener Terms and Their Meaning

English threads as expressed as follows:

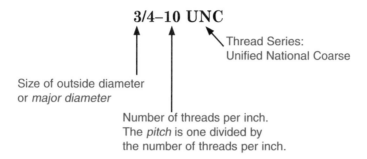

3/4–10 UNC

Thread Series:
Unified National Coarse

Size of outside diameter
or *major diameter*

Number of threads per inch.
The *pitch* is one divided by
the number of threads per inch.

Metric threads are expressed differently:

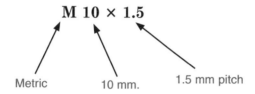

M 10 × 1.5

Metric   10 mm.   1.5 mm pitch

Figure 8.2 shows the pitch of a thread.

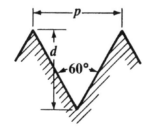

Figure 8.2 The pitch of a thread (P) is one divided by the number
of threads per inch. Both english and metric fasteners
have a 60° thread angle.

SECTION 8

Shop Recommended page 1683, Table 5: Drill and Counterbore Sizes for Socket Head Cap Screws

This is a section view of a counterbored hole. "A" diameter is .012 to .06 larger than "D" (thread diameter). The counterboring tool has a *pilot* diameter that is a slip fit in "A."

Table 5. Drill and Counterbore Sizes for Socket Head Cap Screws (1960 Series)

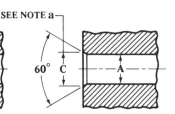

This is the socket head cap screw. "D" is the diameter of the threaded and unthreaded portion of the screw. "F" dimension is a radius for strength.

Nominal Size or Basic Screw Diameter		Nominal Drill Size				Counterbore Diameter	Countersink Diameter[a]
		Close Fit[b]		Normal Fit[c]			
		Number or Fractional Size	Decimal Size	Number or Fractional Size	Decimal Size		
		A				B	C
0	0.0600	51	0.067	49	0.073	1/8	0.074
1	0.0730	46	0.081	43	0.089	5/32	0.087
2	0.0860	3/32	0.094	36	0.106	3/16	0.102
3	0.0990	36	0.106	31	0.120	7/32	0.115
4	0.1120	1/8	0.125	29	0.136	7/32	0.130
5	0.1250	9/64	0.141	23	0.154	1/4	0.145
6	0.1380	23	0.154	18	0.170	9/32	0.158
8	0.1640	15	0.180	10	0.194	5/16	0.188
10	0.1900	5	0.206	2	0.221	3/8	0.218
1/4	0.2500	17/64	0.266	9/32	0.281	7/16	0.278
5/16	0.3125	21/64	0.328	11/32	0.344	17/32	0.346
3/8	0.3750	25/64	0.391	13/32	0.406	5/8	0.415
7/16	0.4375	29/64	0.453	15/32	0.469	23/32	0.483
1/2	0.5000	33/64	0.516	17/32	0.531	13/16	0.552
5/8	0.6250	41/64	0.641	21/32	0.656	1	0.689
3/4	0.7500	49/64	0.766	25/32	0.781	1 3/16	0.828
7/8	0.8750	57/64	0.891	29/32	0.906	1 3/8	0.963

Fasteners smaller than 1/4″ are expressed as numbers

To make this simple, the clearance hole for small fasteners (expressed as numbers) can be .015.

Clearance holes for fasteners from 1/4″ to 7/16″ diameter can be 1/32″ (.03).

Clearance holes for fasteners larger than 1/2″ diameter can be 1/16″ (.06).

170

Fasteners

Always check the pilot diameter of the counterbore before drilling to make sure that it will fit into the clearance hole!

Analyze, Evaluate, & Implement

Make a table listing popular sizes of socket head cap screws. For every size of socket head cap screw, list the size of the clearance drill, the diameter of the screw head, and the recommended depth of the counterbore. The counterbore depth should allow the fastener to be .03 recessed below the surface of the material. Save the table for reference.

Shop Recommended page 1684, Table 6: American National Hexagon and Spline Socket Flat Head Cap Screws

1684 CAP SCREWS

Table 6. American National Standard Hexagon and Spline Socket Flat Countersunk Head Cap Screws *ANSI/ASME B18.3-1998*

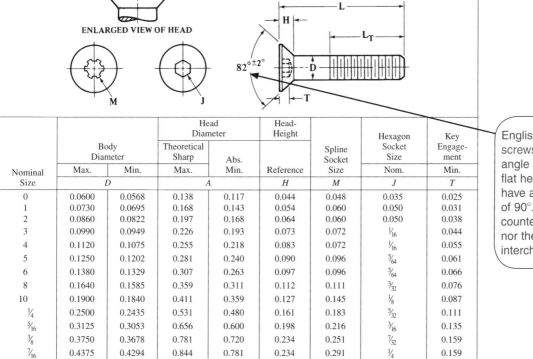

THEORETICAL SHARP — ABSOLUTE MINIMUM — ROUND OR FLAT — ENLARGED VIEW OF HEAD — M — J — 82°±2° — L — H — L_T — D — T

English flat head cap screws have a head angle of 82°. Metric flat head cap screws have a head angle of 90°. Neither the countersinking tool nor the fasteners are interchangeable.

Nominal Size	Body Diameter Max.	Body Diameter Min.	Head Diameter Theoretical Sharp Max.	Head Diameter Abs. Min.	Head-Height Reference	Spline Socket Size	Hexagon Socket Size Nom.	Key Engagement Min.
	D		*A*		*H*	*M*	*J*	*T*
0	0.0600	0.0568	0.138	0.117	0.044	0.048	0.035	0.025
1	0.0730	0.0695	0.168	0.143	0.054	0.060	0.050	0.031
2	0.0860	0.0822	0.197	0.168	0.064	0.060	0.050	0.038
3	0.0990	0.0949	0.226	0.193	0.073	0.072	1/16	0.044
4	0.1120	0.1075	0.255	0.218	0.083	0.072	1/16	0.055
5	0.1250	0.1202	0.281	0.240	0.090	0.096	5/64	0.061
6	0.1380	0.1329	0.307	0.263	0.097	0.096	5/64	0.066
8	0.1640	0.1585	0.359	0.311	0.112	0.111	3/32	0.076
10	0.1900	0.1840	0.411	0.359	0.127	0.145	1/8	0.087
1/4	0.2500	0.2435	0.531	0.480	0.161	0.183	5/32	0.111
5/16	0.3125	0.3053	0.656	0.600	0.198	0.216	3/16	0.135
3/8	0.3750	0.3678	0.781	0.720	0.234	0.251	7/32	0.159
7/16	0.4375	0.4294	0.844	0.781	0.234	0.291	1/4	0.159
1/2	0.5000	0.4919	0.938	0.872	0.251	0.372	5/16	0.172
5/8	0.6250	0.6163	1.188	1.112	0.324	0.454	3/8	0.220

8

171

Flat Head Cap Screws

Flat head cap screws have a tapered head that mate with a *countersunk* hole. *Countersinking* is a machining operation that is used to recess a flathead fastener below the surface of the material or to provide an angle or a lead on a hole. To break a sharp corner, *countersink* a hole to provide a chamfer of .03 to .06 or 10% of the hole diameter per side.

The decision to use flat head cap screws should not be taken lightly. This type of fastener positively locks components together by the mating angles between the fastener and the part. If this type of application is desirable, take care in machining the hole locations and countersinks. Flat head fasteners should always be recessed below the surface of the part. A small locational error will cause the head of the fastener to protrude above the part.

Set Screws

Set screws have an external thread and are driven by a screw driver or a hex key. They are used to lock collars on shafts, hold cutting tools in tool holders, and applications where there is not room for other types of fasteners.

Socket head set screws are abbreviated on engineering drawings as "SHSS." Different point styles are available:

- Cup
- Cone
- Flat
- Oval
- Half dog

8

Fasteners

Shop Recommended page 1693, Table 15: Hexagon and Spline Socket Set Screws

Table 15. ANSI Hexagon and Spline Socket Set Screws *ANSI/ASME B18.3-1998*

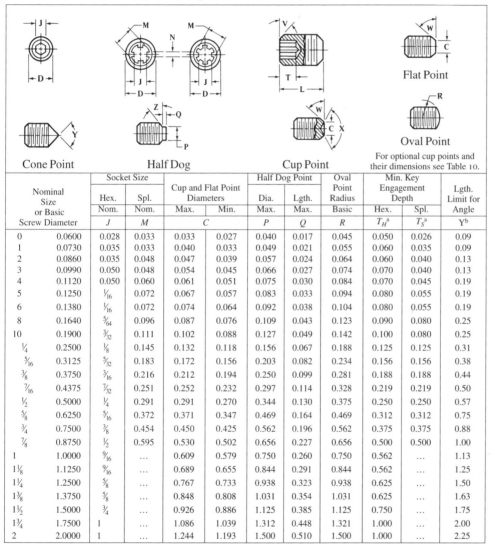

Cone Point   Half Dog   Cup Point

Flat Point

Oval Point

For optional cup points and their dimensions see Table 10.

Different point styles of set screws

Nominal Size or Basic Screw Diameter		Socket Size		Cup and Flat Point Diameters		Half Dog Point		Oval Point Radius	Min. Key Engagement Depth		Lgth. Limit for Angle
		Hex. Nom.	Spl. Nom.	Max.	Min.	Dia. Max.	Lgth. Max.	Basic	Hex.	Spl.	
		J	M	C		P	Q	R	$T_H{}^a$	$T_S{}^a$	Y^b
0	0.0600	0.028	0.033	0.033	0.027	0.040	0.017	0.045	0.050	0.026	0.09
1	0.0730	0.035	0.033	0.040	0.033	0.049	0.021	0.055	0.060	0.035	0.09
2	0.0860	0.035	0.048	0.047	0.039	0.057	0.024	0.064	0.060	0.040	0.13
3	0.0990	0.050	0.048	0.054	0.045	0.066	0.027	0.074	0.070	0.040	0.13
4	0.1120	0.050	0.060	0.061	0.051	0.075	0.030	0.084	0.070	0.045	0.19
5	0.1250	1/16	0.072	0.067	0.057	0.083	0.033	0.094	0.080	0.055	0.19
6	0.1380	1/16	0.072	0.074	0.064	0.092	0.038	0.104	0.080	0.055	0.19
8	0.1640	5/64	0.096	0.087	0.076	0.109	0.043	0.123	0.090	0.080	0.25
10	0.1900	3/32	0.111	0.102	0.088	0.127	0.049	0.142	0.100	0.080	0.25
1/4	0.2500	1/8	0.145	0.132	0.118	0.156	0.067	0.188	0.125	0.125	0.31
5/16	0.3125	5/32	0.183	0.172	0.156	0.203	0.082	0.234	0.156	0.156	0.38
3/8	0.3750	3/16	0.216	0.212	0.194	0.250	0.099	0.281	0.188	0.188	0.44
7/16	0.4375	7/32	0.251	0.252	0.232	0.297	0.114	0.328	0.219	0.219	0.50
1/2	0.5000	1/4	0.291	0.291	0.270	0.344	0.130	0.375	0.250	0.250	0.57
5/8	0.6250	5/16	0.372	0.371	0.347	0.469	0.164	0.469	0.312	0.312	0.75
3/4	0.7500	3/8	0.454	0.450	0.425	0.562	0.196	0.562	0.375	0.375	0.88
7/8	0.8750	1/2	0.595	0.530	0.502	0.656	0.227	0.656	0.500	0.500	1.00
1	1.0000	9/16	…	0.609	0.579	0.750	0.260	0.750	0.562	…	1.13
1 1/8	1.1250	9/16	…	0.689	0.655	0.844	0.291	0.844	0.562	…	1.25
1 1/4	1.2500	5/8	…	0.767	0.733	0.938	0.323	0.938	0.625	…	1.50
1 3/8	1.3750	5/8	…	0.848	0.808	1.031	0.354	1.031	0.625	…	1.63
1 1/2	1.5000	3/4	…	0.926	0.886	1.125	0.385	1.125	0.750	…	1.75
1 3/4	1.7500	1	…	1.086	1.039	1.312	0.448	1.321	1.000	…	2.00
2	2.0000	1	…	1.244	1.193	1.500	0.510	1.500	1.000	…	2.25

8

SECTION 8

Different point styles make different types of indentations in the mating part.

- A cup point makes a circular impression in the mating part to help hold it in position.
- A half dog has a smaller diameter that fits into a drilled hole of the same diameter in the mating part.
- Cone and oval points reproduce the shape of the point in the part and have better holding power.
- Flat points set screws are used when indentations in the part are undesirable.
- Large set screws with a square head are typically used on machine tools for leveling pads.

Socket Head Shoulder Screws (Shoulder Bolts or Stripper Bolts)

 Index navigation path and key words:

— shoulder screws/page 1686

Shoulder bolts have a hex head, a straight ground diameter, and a threaded end. The ground diameter is a slip fit with the mating component. These fasteners are sometimes called "stripper bolts" because they are used on stamping dies to contain and limit the stroke of mating components (see Figure 8.3). The ground diameter can also be a slip fit into a bushing for a part that rotates (see Figure 8.4).

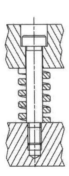 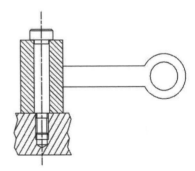

Figure 8.3 Shoulder bolt used to contain a spring in a stamping die.

Figure 8.4 Shoulder bolt used as a pivot point for a handle.

Fasteners

Shop Recommended page 1686, Table 8

1686   CAP SCREWS

Table 8. American National Standard Hexagon Socket Head Shoulder Screws
ANSI/ASME B18.3-1998

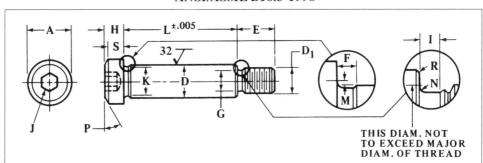

THIS DIAM. NOT TO EXCEED MAJOR DIAM. OF THREAD

> Note that diameter "D" is slightly less than *nominal* size for a slip fit with the mating component.

Nominal Size	Shoulder Diameter Max.	Shoulder Diameter Min.	Head Diameter Max.	Head Diameter Min.	Head Height Max.	Head Height Min.	Head Side Height Min.	Nominal Thread Size	Thread Length
	D		A		H		S	D_1	E
¼	0.2480	0.2460	0.375	0.357	0.188	0.177	0.157	10-24	0.375
5/16	0.3105	0.3085	0.438	0.419	0.219	0.209	0.183	¼-20	0.438
⅜	0.3730	0.3710	0.562	0.543	0.250	0.240	0.209	5/16-18	0.500
½	0.4980	0.4960	0.750	0.729	0.312	0.302	0.262	⅜-16	0.625
⅝	0.6230	0.6210	0.875	0.853	0.375	0.365	0.315	½-13	0.750
¾	0.7480	0.7460	1.000	0.977	0.500	0.490	0.421	⅝-11	0.875
1	0.9980	0.9960	1.312	1.287	0.625	0.610	0.527	¾-10	1.000
1¼	1.2480	1.2460	1.750	1.723	0.750	0.735	0.633	⅞-9	1.125
1½	1.4980	1.4960	2.125	2.095	1.000	0.980	0.842	1⅛-7	1.500
1¾	1.7480	1.7460	2.375	2.345	1.125	1.105	0.948	1¼-7	1.750
2	1.9980	1.9960	2.750	2.720	1.250	1.230	1.054	1½-6	2.000

Nominal Size	Thread Neck Diameter Max.	Thread Neck Diameter Min.	Thread Neck Width Max.	Shoulder Neck Dia. Min.	Shoulder Neck Width Max.	Thread Neck Fillet Max.	Thread Neck Fillet Min.	Head Fillet Extension Above D Max.	Hexagon Socket Size Nom.
	G		I	K	F	N		M	J
¼	0.142	0.133	0.083	0.227	0.093	0.023	0.017	0.014	⅛
5/16	0.193	0.182	0.100	0.289	0.093	0.028	0.022	0.017	5/32
⅜	0.249	0.237	0.111	0.352	0.093	0.031	0.025	0.020	3/16
½	0.304	0.291	0.125	0.477	0.093	0.035	0.029	0.026	¼
⅝	0.414	0.397	0.154	0.602	0.093	0.042	0.036	0.032	5/16
¾	0.521	0.502	0.182	0.727	0.093	0.051	0.045	0.039	⅜
1	0.638	0.616	0.200	0.977	0.125	0.055	0.049	0.050	½
1¼	0.750	0.726	0.222	1.227	0.125	0.062	0.056	0.060	⅝
1½	0.964	0.934	0.286	1.478	0.125	0.072	0.066	0.070	⅞
1¾	1.089	1.059	0.286	1.728	0.125	0.072	0.066	0.080	1
2	1.307	1.277	0.333	1.978	0.125	0.102	0.096	0.090	1¼

8

175

When tapping a hole for a shoulder screw, do not machine a deep chamfer that will cause the ground diameter of the shoulder bolt to enter the chamfer. The face of the shoulder bolt must rest on the flat surface of the plate as shown in Figure 8.5

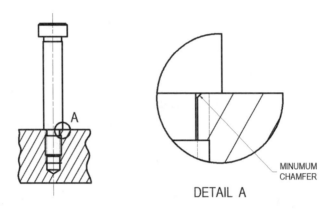

DETAIL A

MINUMUM
CHAMFER

Figure 8.5 Shoulder portion of a shoulder bolt must sit flat on mounting surface.

Unit K: Pins and Studs pages 1746–1762

Dowel Pins

Hardened and ground dowel pins are made to a high degree of accuracy. Dowel pins are used to:

- Position and align machine components
- Provide a positive stop
- Add structural rigidity to an assembly
- Create a part nest

When components are assembled with dowel pins, the dowel is a very close fit with its mating part. Machines or tools made with dowel pin construction can be disassembled for maintenance, engineering changes, or repair, and reassembled with the same positional accuracy. Holes for dowel pins are carefully produced to obtain either a close sliding fit or a press fit for the dowel. Dowel pins are used in assemblies with socket head cap screws. The screw holds the components together; the dowel pin locates the components. When using dowels with fasteners, proper machine shop practice dictates that the sizes should be the same whenever possible. For example, if 3/8 socket head cap screws are used, 3/8 diameter dowels should be used also.

Holes for dowel pins are finished with a chucking reamer to provide the type fit desired. Reamers come in *nominal* sizes, which are standard sizes from which tolerances and allowances are established. For example, when used properly, a .3750 diameter reamer finishes a hole nearly exactly the size of the reamer for a light press fit. A press fit is when the dowel is larger than the hole in the mating part.

This end of the dowel has an angular lead for easier insertion into the hole.

Use a soft mallet, a drift, or an arbor press to press the dowel into the hole.

"L" dimension is the length of the precisely ground diameter. Ideal amount of engagement is two times the dowel diameter.

Oversize dowels are available for repairs, but they are not used routinely because they are not considered an "off the shelf" standard part. As a result they cannot be replaced with stock hardware.

Pin diameter "A" is .0002 to .0003 larger than nominal. When the mating hole is produced to "size," the press fit is correct.

Table 1. American National Standard Hardened Ground Machine Dowel Pins *ANSI/ASME B18.8.2-1995*

Nominal Size[a] or Nominal Pin Diameter		Pin Diameter, A						Point Diameter, B		Crown Height, C	Crown Radius, R	Range of Preferred Lengths,[b] L	Single Shear Load, for Carbon or Alloy Steel, Calculated lb	Suggested Hole Diameter[c]	
		Standard Series Pins			Oversize Series Pins										
		Basic	Max	Min	Basic	Max	Min	Max	Min	Max	Min			Max	Min
1/16	0.0625	0.0627	0.0628	0.0626	0.0635	0.0636	0.0634	0.058	0.048	0.020	0.008	3/16-3/4	400	0.0625	0.0620
5/64 [d]	0.0781	0.0783	0.0784	0.0782	0.0791	0.0792	0.0790	0.074	0.064	0.026	0.010	...	620	0.0781	0.0776
3/32	0.0938	0.0940	0.0941	0.0939	0.0948	0.0949	0.0947	0.089	0.079	0.031	0.012	5/16-1	900	0.0937	0.0932
1/8	0.1250	0.1252	0.1253	0.1251	0.1260	0.1261	0.1259	0.120	0.110	0.041	0.016	3/8-2	1,600	0.1250	0.1245
5/32 [d]	0.1562	0.1564	0.1565	0.1563	0.1572	0.1573	0.1571	0.150	0.140	0.052	0.020	...	2,500	0.1562	0.1557
3/16	0.1875	0.1877	0.1878	0.1876	0.1885	0.1886	0.1884	0.180	0.170	0.062	0.023	1/2-2	3,600	0.1875	0.1870
1/4	0.2500	0.2502	0.2503	0.2501	0.2510	0.2511	0.2509	0.240	0.230	0.083	0.031	1/2-2 1/2	6,400	0.2500	0.2495
5/16	0.3125	0.3127	0.3128	0.3126	0.3135	0.3136	0.3134	0.302	0.290	0.104	0.039	1/2-2 1/2	10,000	0.3125	0.3120
3/8	0.3750	0.3752	0.3753	0.3751	0.3760	0.3761	0.3759	0.365	0.350	0.125	0.047	1/2-3	14,350	0.3750	0.3745
7/16	0.4375	0.4377	0.4378	0.4376	0.4385	0.4386	0.4384	0.424	0.409	0.146	0.055	7/8-3	19,550	0.4375	0.4370
1/2	0.5000	0.5002	0.5003	0.5001	0.5010	0.5011	0.5009	0.486	0.471	0.167	0.063	3/4, 1-4	25,500	0.5000	0.4995
5/8	0.6250	0.6252	0.6253	0.6251	0.6260	0.6261	0.6259	0.611	0.595	0.208	0.078	1 1/4-5	39,900	0.6250	0.6245
3/4	0.7500	0.7502	0.7503	0.7501	0.7510	0.7511	0.7509	0.735	0.715	0.250	0.094	1 1/2-6	57,000	0.7500	0.7495
7/8	0.8750	0.8752	0.8753	0.8751	0.8760	0.8761	0.8759	0.860	0.840	0.293	0.109	2, 2 1/2-6	78,000	0.8750	0.8745
1	1.0000	1.0002	1.0003	1.0001	1.0010	1.0011	1.0009	0.980	0.960	0.333	0.125	2, 2 1/2-5.6	102,000	1.0000	0.9995

8

177

Making It Simple

Figure 8.6 is a typical example of an assembly using socket head cap screws and dowels. The fasteners and dowels are positioned as far apart as possible for rigidity and locational stability. Engineering drawings often show dowel locations as a circle with a partly filled-in cross. The dowel holes on this assembly are relieved in the top plate and have a smaller knockout hole in the bottom plate. The hole for the socket head cap screw is counterbored deep enough so that the screw is recessed .03 below the surface of the plate. Making dimension "A" different from dimension "B" prevents the plates from being assembled backward.

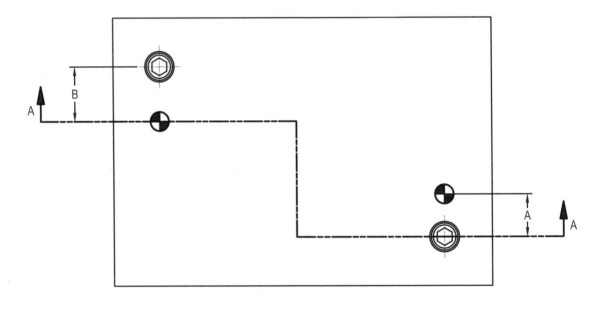

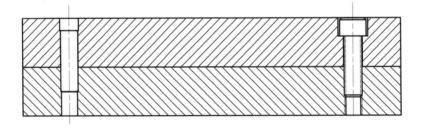

SECTION A-A

Figure 8.6 An assembly using socket head cap screws and dowels.

Tips for when dowels are used for positioning:

- Position dowel pins as far apart as possible.
- Engagement for the dowel should be two times the dowel diameter in each component.
- Never press a dowel into a blind (dead end) hole.
- Do not press a dowel into a hardened part.
- Provide a knock out hole.
- Relieve deep holes.
- Dowel should be a press fit in one component and a slip fit in the other.

Shop tip

When designing assemblies that use dowels, if one component is hardened and one component is soft, the press fit should be in the soft component, and the slip fit should be in the hardened component. Pressing hardened dowels in to hardened parts is not recommended because cracks can occur.

 For more information on *reaming*, see Section 5, Tooling Toolmaking/Reamers, MH29 pages 844–865

Dowels in an Assembly

Figure 8.7 shows a cross section of two plates with four different assembly possibilities for dowels.

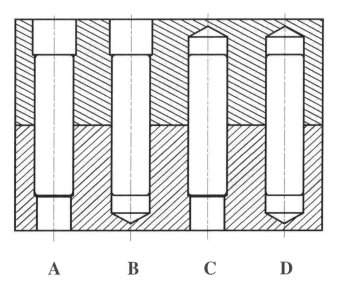

A   B   C   D

Figure 8.7 Assembly possibilities for dowels.

8

Option A

In the lower plate the dowel is a light press fit, which means that there is .0002 to .0005 interference between the dowel and the hole. In other words, the dowel is larger than the hole. The hole in the bottom part of the lower plate is made .03 *smaller* than the dowel hole to prevent the dowel from falling out. This is also a knockout hole for a punch. Always provide a way to disassemble a tool. The top plate in option A has a hole that is .0002 to .0005 *larger* than the dowel for a slip fit. The larger hole above the dowel is a relief. The relief is .03 larger than the dowel. It is a good practice to relieve portions of the hole that are not used for locating. Dowels should be installed with two times their diameter as a bearing surface. In other words, a 3/8″ diameter dowel should contact the hole for 3/4ths of its length in *each* component. A 1/4″dowel should contact each plate for 1/2″ of the dowel's length in each component.

Option B

In this example the design does not permit the dowel hole to go through the lower plate. Because pressing dowels into blind (dead end) holes is not recommended, the dowel is a slip fit in the lower plate and a press fit in the top plate. Again, the top plate is relieved, allowing the dowel the proper amount of effective engagement.

Option C

This example is the opposite of Option B. A knockout hole which is .03 smaller than the dowel is provided in the lower plate. The press fit is in the lower plate, the slip fit is in the top plate because it is very difficult to remove a dowel that is pressed in a blind hole.

Option D

In the unlikely event that through holes are not permitted in either plate, both sides are made to be a slip fit for the dowel.

Spring Pins

Spring pins or "roll pins" are like dowel pins, but are for applications that are less critical. Holes for spring pins are just drilled which is crude compared to the way dowel pin holes are machined. Spring pins are slightly oversize, so a common fractional drill can be used to produce the hole.

Shop Recommended page 1761, Table 10

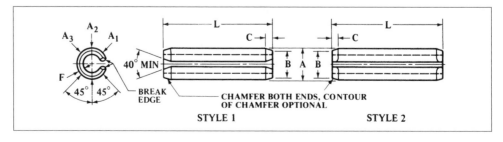

180   *Spring Pin*

Studs

Studs are straight and threaded on both ends. A common application for a stud is in an automobile engine for holding down the cylinder head. When assembling, the studs in the engine block guide the head into place. Studs are also used on machine tools to clamp parts and vises to the machine table.

Unit L: Retaining Rings pages 1763–1790

Retaining rings fit into a machined groove on a shaft or in a groove inside a hole. A common application for a retaining ring is to provide an artificial shoulder for a bearing or bushing. The four most common types are external, internal, E-type, and spiral. Some retaining rings snap onto the outside diameter of a shaft and can be installed without special tools. E type are also called "E-clips" because they look like the letter "E" or "Jesus clips" by the old-timers for reasons unknown.

Shop Recommended pages 1778–1784, Tables 10, 11, 12, 14

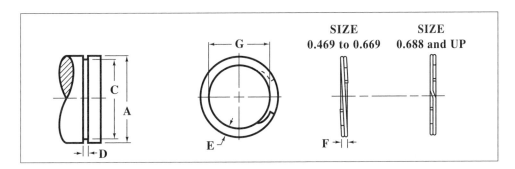

Table 10, Spiral

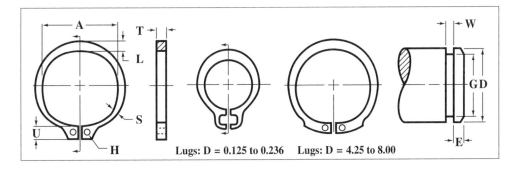

Table 11, External

8

181

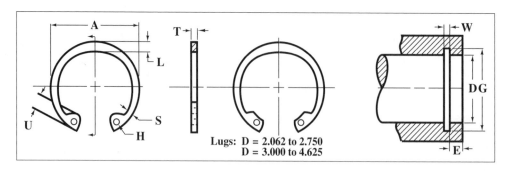

Table 12, Internal

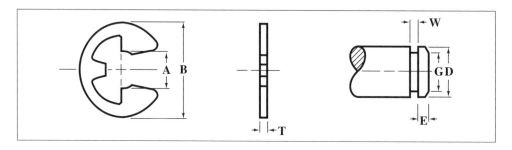

Table 14, E type

 Unit A: Torque and Tension in Fasteners pages 1521–1537

 Index navigation paths and key words:

 Unit B: Inch Threaded Fasteners pages 1538–1579

Unit C: Metric Threaded Fasteners pages 1581–1612

Index navigation paths and key words:

8

Check these pages for information:

BRITISH FASTENERS

British Standard Square and Hexagon Bolts, Screws and Nuts.—Important dimensions of precision hexagon bolts, screws and nuts (BSW and BSF threads) as covered by British Standard 1083:1965 are given in Tables 1 and 2. The use of fasteners in this standard will decrease as fasteners having Unified inch and ISO metric threads come into increasing use.

Index navigation paths and key words:

8

SECTION 8

ASSIGNMENT

List the key terms and give a definition of each.

Nominal	Bolt
Blind Hole	Screw
Press Fit	Pilot Diameter
Slip Fit	Counterbore
Hexagon Head	Countersink

APPLY IT!

1. What is the difference between a screw and a bolt?

2. Flat head cap screws are used for what types of applications?

3. What length of engagement is recommended for the following dowel sizes?
   a. 1/2″
   b. .3750
   c. 1/4″
   d. .1875

4. What is the minimum thread engagement for the following thread sizes?
   a. 10–32
   b. .50–13
   c. 3/8–16
   d. 3/8–24

5. Why are most dowels made to a length that is four times its diameter?

6. How deep is the counterbore for the following socket head cap screws? Add .03 to the depth for head clearance.
   a. 3/8–16 × 1.5 SHCS
   b. 1/2–13 × 1.0 SHCS
   c. 1/4–20 × .75 SHCS
   d. 10–32 × 3/8 SHCS

8

184

Section 9

THREADS and THREADING

Use this section with pages 1802 — 2121
in Machinery's Handbook 29th Edition

SECTION 9

Navigation Overview

Units Covered in this Section with:
Navigation Assistant

The Navigation Assistant helps find information in the MH29 Primary Index. The Primary Index is located in the back of the book on pages 2701–2788 and is set up alphabetically by subject. Watch for the magnifying glass throughout the section for navigation hints.

9

Key Terms

Chamfer	Major Diameter
Crest	Thread Series
Unified National Coarse	Nominal Size
Unified National Fine	Pitch Diameter
Minor Diameter	Root

Learning Objectives

After studying this unit you should be able to:

- Identify the key threading terms used in manufacturing.
- Recognize the most common thread forms used in industry.
- Measure threads by three different methods.
- Determine tap drill sizes.
- Determine tap drill size based on percentage of thread.

Introduction

Producing threads is one of the most common machine shop operations. Nuts are rarely used on special machines, tools, dies, or fixtures. Internal threads are produced by tapping because they are much stronger and better for machine design. External threads are can be cut on a lathe or with a die. Threads and fasteners are standardized. Compliance is overseen by organizations such as the American National Standards Institute (ANSI).

Unit A: Screw Thread Systems pages 1806–1812

After World War II, the North Atlantic Treaty Organization adopted the *Unified Thread System* as a standard thread form. *Unified threads* include the following series: UNC (Unified National Coarse) and UNF (Unified National Fine). Metric threads are expressed in millimeters. The most widely used thread forms that are used for fastening have a 60-degree included thread angle. The *pitch* (P) of a thread is the distance from one point of the thread to the corresponding point of the next thread. The thread depth (d) is the distance from the *crest* (*major diameter*) of the thread to *root* (*minor diameter*). The *pitch diameter* of a thread is an imaginary cylinder halfway between the *crest* and *root*.

9

SECTION 9

Index navigation paths and key words:

— unified thread system/screw thread form/pages 1806–1808, 1812–1864

Inch threads include UNC (Unified National Coarse) and UNF (Unified National Fine). Inch thread systems identify the thread size by the outside (major diameter) followed by the number of threads in one inch. Inch threads have different *classes* depending on their application. A *class 2* thread is the most popular because it is used for most fasteners. Class 1 threads are a loose fit for applications that require greater freedom of assembly. Class 3 threads provide a closer fit. Left-handed threads are shown as LH after the thread designation. Unified threads smaller than 1/4″ are expressed in numbers 0–12. The formula for converting number system threads to decimal inch is: thread number $\times$ .013 + .06 = decimal inch equivalent. For example, a # 10 fastener is .190 because 10 $\times$ .013 + .06 = .190. A 6-32 fastener is 6 $\times$ .013 + .06 =.138.

English threads are expressed as follows:

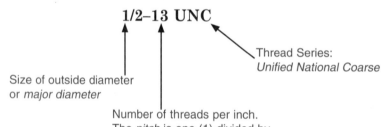

1/2–13 UNC

Size of outside diameter or *major diameter*

Number of threads per inch. The *pitch* is one (1) divided by the number of threads per inch. The *pitch* of this thread is .0769

Thread Series: *Unified National Coarse*

9

Shop Recommended page 1807 Basic Profile of UN and UNF Screw Threads

Accurate measurements can be taken on the Pitch Diameter, which Is the functional diameter of the thread

The *pitch*(P) of a thread is the distance from one point of a thread to the corresponding point of the next thread.

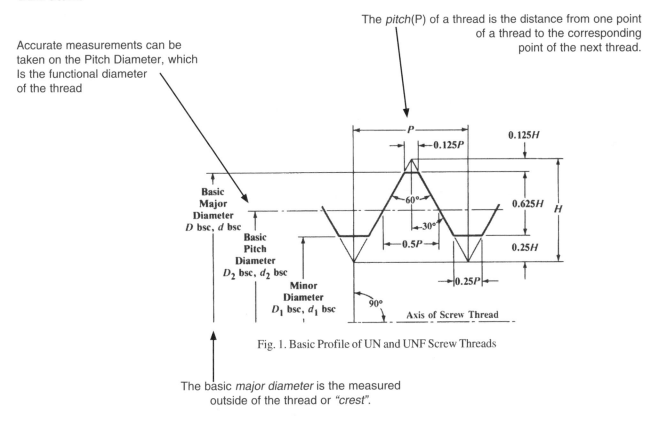

Fig. 1. Basic Profile of UN and UNF Screw Threads

The basic *major diameter* is the measured outside of the thread or *"crest"*.

Shop Recommended page 1817, Table 3

Table 3 "Standard Series and Selected Combinations – Unified Screw Threads" is useful in the manufacturing and inspection of screw threads. The *nominal size* of the thread is listed in the left hand column of the table along with the number of threads per inch and the *series* of the thread. *Nominal size* is the size used for the general identification of the thread. The *nominal* size of a 3/8–16 thread is 3/8. The *series* of a thread is the standard that controls all aspects of the thread. The *Major Diameter* of an external thread is its outside diameter. This is the measured outside of the thread and is slightly less than the *nominal* size. The *major diameter* is also known as the *crest.* The *root* is the bottom of an external thread and is also known as the *minor diameter.*

9

Table 3. Standard Series and Selected Combinations — Unified Screw Threads

Nominal Size, Threads per Inch, and Series Designation[a]	External[b]								Internal[b]					
			Major Diameter			Pitch Diameter		UNR Minor Dia.[c] Max (Ref.)		Minor Diameter		Pitch Diameter		Major Diameter
	Class	Allow-ance	Max[d]	Min	Min[e]	Max[d]	Min		Class	Min	Max	Min	Max	Min
0-80 UNF	2A	0.0005	0.0595	0.0563	—	0.0514	0.0496	0.0446	2B	0.0465	0.0514	0.0519	0.0542	0.0600
	3A	0.0000	0.0600	0.0568	—	0.0519	0.0506	0.0451	3B	0.0465	0.0514	0.0519	0.0536	0.0600
1-64 UNC	2A	0.0006	0.0724	0.0686	—	0.0623	0.0603	0.0538	2B	0.0561	0.0623	0.0629	0.0655	0.0730
	3A	0.0000	0.0730	0.0692	—	0.0629	0.0614	0.0544	3B	0.0561	0.0623	0.0629	0.0648	0.0730
1-72 UNF	2A	0.0006	0.0724	0.0689	—	0.0634	0.0615	0.0559	2B	0.0580	0.0635	0.0640	0.0665	0.0730
	3A	0.0000	0.0730	0.0695	—	0.0640	0.0626	0.0565	3B	0.0580	0.0635	0.0640	0.0659	0.0730
2-56 UNC	2A	0.0006	0.0854	0.0813	—	0.0738	0.0717	0.0642	2B	0.0667	0.0737	0.0744	0.0772	0.0860
	3A	0.0000	0.0860	0.0819	—	0.0744	0.0728	0.0648	3B	0.0667	0.0737	0.0744	0.0765	0.0860
2-64 UNF	2A	0.0006	0.0854	0.0816	—	0.0753	0.0733	0.0668	2B	0.0691	0.0753	0.0759	0.0786	0.0860
	3A	0.0000	0.0860	0.0822	—	0.0759	0.0744	0.0674	3B	0.0691	0.0753	0.0759	0.0779	0.0860
3-48 UNC	2A	0.0007	0.0983	0.0938	—	0.0848	0.0825	0.0734	2B	0.0764	0.0845	0.0855	0.0885	0.0990
	3A	0.0000	0.0990	0.0945	—	0.0855	0.0838	0.0741	3B	0.0764	0.0845	0.0855	0.0877	0.0990
3-56 UNF	2A	0.0007	0.0983	0.0942	—	0.0867	0.0845	0.0771	2B	0.0797	0.0865	0.0874	0.0902	0.0990
	3A	0.0000	0.0990	0.0949	—	0.0874	0.0858	0.0778	3B	0.0797	0.0865	0.0874	0.0895	0.0990
4-40 UNC	2A	0.0008	0.1112	0.1061	—	0.0950	0.0925	0.0814	2B	0.0849	0.0939	0.0958	0.0991	0.1120
	3A	0.0000	0.1120	0.1069	—	0.0958	0.0939	0.0822	3B	0.0849	0.0939	0.0958	0.0982	0.1120
4-48 UNF	2A	0.0007	0.1113	0.1068	—	0.0978	0.0954	0.0864	2B	0.0894	0.0968	0.0985	0.1016	0.1120
	3A	0.0000	0.1120	0.1075	—	0.0985	0.0967	0.0871	3B	0.0894	0.0968	0.0985	0.1008	0.1120
5-40 UNC	2A	0.0008	0.1242	0.1191	—	0.1080	0.1054	0.0944	2B	0.0979	0.1062	0.1088	0.1121	0.1250
	3A	0.0000	0.1250	0.1199	—	0.1088	0.1069	0.0952	3B	0.0979	0.1062	0.1088	0.1113	0.1250
5-44 UNF	2A	0.0007	0.1243	0.1195	—	0.1095	0.1070	0.0972	2B	0.1004	0.1079	0.1102	0.1134	0.1250
	3A	0.0000	0.1250	0.1202	—	0.1102	0.1083	0.0979	3B	0.1004	0.1079	0.1102	0.1126	0.1250
6-32 UNC	2A	0.0008	0.1372	0.1312	—	0.1169	0.1141	0.1000	2B	0.104	0.114	0.1177	0.1214	0.1380
	3A	0.0000	0.1380	0.1320	—	0.1177	0.1156	0.1008	3B	0.1040	0.1140	0.1177	0.1204	0.1380
6-40 UNF	2A	0.0008	0.1372	0.1321	—	0.1210	0.1184	0.1074	2B	0.111	0.119	0.1218	0.1252	0.1380
	3A	0.0000	0.1380	0.1329	—	0.1218	0.1198	0.1082	3B	0.1110	0.1186	0.1218	0.1243	0.1380
8-32 UNC	2A	0.0009	0.1631	0.1571	—	0.1428	0.1399	0.1259	2B	0.130	0.139	0.1437	0.1475	0.1640
	3A	0.0000	0.1640	0.1580	—	0.1437	0.1415	0.1268	3B	0.1300	0.1389	0.1437	0.1465	0.1640
8-36 UNF	2A	0.0008	0.1632	0.1577	—	0.1452	0.1424	0.1301	2B	0.134	0.142	0.1460	0.1496	0.1640
	3A	0.0000	0.1640	0.1585	—	0.1460	0.1439	0.1309	3B	0.1340	0.1416	0.1460	0.1487	0.1640

9

Size of thread; number of threads in one inch; and series of thread. Thread sizes smaller than 1/4 are expressed as numbers 0 through 12

Class of thread: Class 1 is a loose fit; Class 2 is a close fit; Class 3 is a free fit; "A" indicates an external thread; "B" indicates an internal thread.

The Pitch Diameter is the diameter of an imaginary cylinder halfway between the root of the thread and the crest; sometimes called the functional diameter.

The Internal portion of Table 3 gives the same type of information as External except that it refers to the female or "inside" threads.

190

Unit D: Metric Screw Threads pages 1878–1920

Metric threads share characteristics with unified threads but they are not interchangeable. They both have 60-degree thread angles and terms regarding the thread are the same such as crest, root, and pitch. Metric threads are expressed in millimeters. The first character in a metric thread designation is "M" for metric. The next number is the size of the major diameter in millimeters. The next number is the *pitch.* If there are numbers and letters following the pitch they refer to the class of fit. A 6g 6h fit is about the same type of fit as a class 2 fit in the decimal inch system. If the metric thread designation does not give the pitch, *always assume it is coarse.* A 1.25 pitch is coarser than a 1.00 pitch because there is more space between threads.

Metric thread designation:

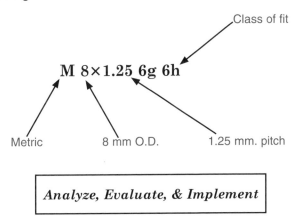

Class of fit

M 8×1.25 6g 6h

Metric   8 mm O.D.   1.25 mm. pitch

| *Analyze, Evaluate, & Implement* |

Use the MH to identify the coarse pitch thread of the following metric thread sizes:

- M6
- M10
- M8

Unit H: Pipe and Hose Threads pages 1956–1972

Pipe threads produce a pressure-tight joint. American National Standard pipe threads provide tapered and straight pipe threads using the following identification system:

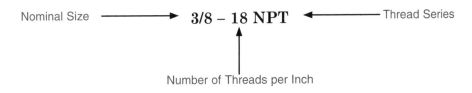

Nominal Size ⟶ **3/8 – 18 NPT** ⟵ Thread Series

Number of Threads per Inch

SECTION 9

Shop Recommended page 1964, Table 8

Table 8. Suggested Tap Drill Sizes for Internal Dryseal Pipe Threads

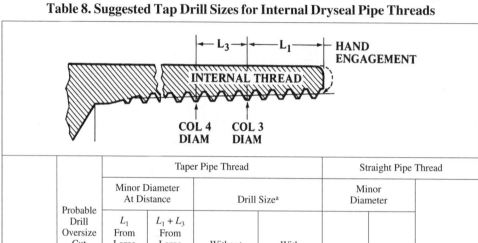

| Size | Probable Drill Oversize Cut (Mean) | Taper Pipe Thread | | | | Straight Pipe Thread | | |
| | | Minor Diameter At Distance | | Drill Size[a] | | Minor Diameter | | |
		L_1 From Large End	$L_1 + L_3$ From Large End	Without Reamer	With Reamer	NPSF	NPSI	Drill Size[a]
$\frac{1}{16}$-27	0.0038	0.2443	0.2374	"C" (0.242)	"A" (0.234)	0.2482	0.2505	"D" (0.246)
$\frac{1}{8}$-27	0.0044	0.3367	0.3298	"Q" (0.332)	$\frac{21}{64}$ (0.328)	0.3406	0.3429	"R" (0.339)
$\frac{1}{4}$-18	0.0047	0.4362	0.4258	$\frac{7}{16}$ (0.438)	$\frac{27}{64}$ (0.422)	0.4422	0.4457	$\frac{7}{16}$ (0.438)
$\frac{3}{8}$-18	0.0049	0.5708	0.5604	$\frac{9}{16}$ (0.562)	$\frac{9}{16}$ (0.563)	0.5776	0.5811	$\frac{37}{64}$ (0.578)
$\frac{1}{2}$-14	0.0051	0.7034	0.6901	$\frac{45}{64}$ (0.703)	$\frac{11}{16}$ (0.688)	0.7133	0.7180	$\frac{45}{64}$ (0.703)
$\frac{3}{4}$-14	0.0060	0.9127	0.8993	$\frac{29}{32}$ (0.906)	$\frac{57}{64}$ (0.891)	0.9238	0.9283	$\frac{59}{64}$ (0.922)
1-11$\frac{1}{2}$	0.0080	1.1470	1.1307	$1\frac{9}{64}$ (1.141)	$1\frac{1}{8}$ (1.125)	1.1600	1.1655	$1\frac{5}{32}$ (1.156)
1$\frac{1}{4}$-11$\frac{1}{2}$	0.0100	1.4905	1.4742	$1\frac{31}{64}$ (1.484)	$1\frac{15}{32}$ (1.469)	…	…	…
1$\frac{1}{2}$-11$\frac{1}{2}$	0.0120	1.7295	1.7132	$1\frac{23}{32}$ (1.719)	$1\frac{45}{64}$ (1.703)	…	…	…
2-11$\frac{1}{2}$	0.0160	2.2024	2.1861	$2\frac{3}{16}$ (2.188)	$2\frac{11}{64}$ (2.172)	…	…	…
2$\frac{1}{2}$-8	0.0180	2.6234	2.6000	$2\frac{39}{64}$ (2.609)	$2\frac{37}{64}$ (2.578)	…	…	…
3-8	0.0200	3.2445	3.2211	$3\frac{15}{64}$ (3.234)	$3\frac{13}{64}$ (3.203)	…	…	…

Size of fitting used, (not the tap diameter)

Drills drill holes that are slightly oversize

Tap drill size for tapered pipe threads. To improve the quality of the pipe threads, drill the hole 0.015 undersize and finish the hole with a reamer.

Tap drill size for straight pipe threads

Unit J: Measuring Screw Threads pages 1989–2014

Index navigation paths and key words:

— threads and threading/measuring screw threads/pages 1989–2014

A variety of methods can be used to measure screw threads, including:

- Thread Micrometers
- Thread Gages
- Three-wire Method

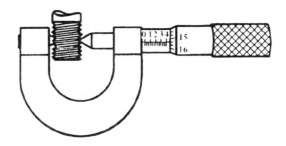

Figure 9.1 Thread micrometer

Thread micrometers have special v-shaped anvils for screw thread measurement.

Thread gages are made to thread onto the part and contact the threads on critical surfaces. There are normally two gages, a "go" gage that must fit on the part and a "no go" gage that cannot fit.

An accurate method of measuring screw threads is to use thread wires. The size of the wire (W) is a diameter that will contact the face of the thread at the *pitch diameter*. Two wires are placed on one side of the thread one wire placed on the opposite side. The measurement (M) is taken over the wires with a micrometer as shown in Figure 9.2. Heavy grease or rubber bands are sometimes used to hold wires in place during measuring. Some tool makers swear by these techniques, others swear at them.

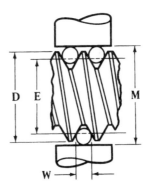

Figure 9.2 Measuring screw threads.

SECTION 9

Apply the reading to the chart on page 1997. The reading over the wires agrees with the dimension shown on the chart, the thread is acceptable.

For example, a 3/8–16 thread uses a wire size of 0.040. The dimension over the three wires should be 0.4003.

Shop Recommended page 1997

Dimensions Over Wires of Given Diameter for Checking Screw Threads of American National Form (U.S. Standard) and the V-Form

Dia. of Thread	No. of Threads per Inch	Wire Dia. Used	Dimension over Wires		Dia. of Thread	No. of Threads per Inch	Wire Dia. Used	Dimension over Wires	
			V-Thread	U.S. Thread				V-Thread	U.S. Thread
1/4	18	0.035	0.2588	0.2708	7/8	8	0.090	0.9285	0.9556
1/4	20	0.035	0.2684	0.2792	7/8	9	0.090	0.9525	0.9766
1/4	22	0.035	0.2763	0.2861	7/8	10	0.090	0.9718	0.9935
1/4	24	0.035	0.2828	0.2919	15/16	8	0.090	0.9910	1.0181
5/16	18	0.035	0.3213	0.3333	15/16	9	0.090	1.0150	1.0391
5/16	20	0.035	0.3309	0.3417	1	8	0.090	1.0535	1.0806
5/16	22	0.035	0.3388	0.3486	1	9	0.090	1.0775	1.1016
5/16	24	0.035	0.3453	0.3544	1⅛	7	0.090	1.1476	1.1785
3/8	16	0.040	0.3867	0.4003	1¼	7	0.090	1.2726	1.3035
3/8	18	0.040	0.3988	0.4108	1⅜	6	0.150	1.5363	1.5724
3/8	20	0.040	0.4084	0.4192	1½	6	0.150	1.6613	1.6974
7/16	14	0.050	0.4638	0.4793	1⅝	5½	0.150	1.7601	1.7995
7/16	16	0.050	0.4792	0.4928	1¾	5	0.150	1.8536	1.8969
1/2	12	0.050	0.5057	0.5237	1⅞	5	0.150	1.9786	2.0219
1/2	13	0.050	0.5168	0.5334	2	4½	0.150	2.0651	2.1132
1/2	14	0.050	0.5263	0.5418	2¼	4½	0.150	2.3151	2.3632
9/16	12	0.050	0.5682	0.5862	2½	4	0.150	2.5170	2.5711
9/16	14	0.050	0.5888	0.6043	2¾	4	0.150	2.7670	2.28211
5/8	10	0.070	0.6618	0.6835	3	3½	0.200	3.1051	3.1670
5/8	11	0.070	0.6775	0.6972	3¼	3½	0.200	3.3551	3.4170
5/8	12	0.070	0.6907	0.7087	3½	3¼	0.250	3.7171	3.7837
11/16	10	0.070	0.7243	0.7460	3¾	3	0.250	3.9226	3.9948
11/16	11	0.070	0.7400	0.7597	4	3	0.250	4.1726	4.2448
3/4	10	0.070	0.7868	0.8085	4¼	2⅞	0.250	4.3975	4.4729
3/4	11	0.070	0.8025	0.8222	4½	2¾	0.250	4.6202	4.6989
3/4	12	0.070	0.8157	0.8337	4¾	2⅝	0.250	4.8402	4.9227
13/16	9	0.070	0.8300	0.8541	5	2½	0.250	5.0572	5.1438
13/16	10	0.070	0.8493	0.8710	…	…	…	…	…

Thread wires for inspecting a 3/8–16 thread

Thread size and number of threads per inch

Select 3 wires of this diameter

Measurement over thread wires

9

194

Threads and Threading

> *Analyze, Evaluate, & Implement*

Obtain English and metric bolts and screws of unknown size and determine size, thread series and pitch, or threads per inch.

Unit K: Tapping and Thread Cutting pages 2015–2047

Tapping is an operation that cuts internal threads. Taps are tapered at the end with a relief ground at the cutting edges. A typical tap makes four to six revolutions before it produces a full thread. To get the proper amount of thread engagement, the hole size for the tapping operation is critical. Taps are driven by machine or by hand with a special wrench. Most taps used in industry are made from high speed steel which has superior wear characteristics, but is brittle. The biggest reason for tap breakage is starting the tap crooked. A tap will not follow a drilled hole. Tools that are made for the removal of broken taps are ineffective at best. Broken taps are most effectively removed by a machine called a tap blaster which works on the same principle as an electrical discharge machine. For adequate strength, the depth of usable threads should be 1 1/2 to 2 times the diameter of the thread.

9

Shop Recommended page 2029, Table 3 Tap Drill sizes for Threads of American National Form

Table 3. Tap Drill Sizes for Threads of American National Form

Screw Thread		Commercial Tap Drills[a]		Screw Thread		Commercial Tap Drills[a]	
Outside Diam. Pitch	Root Diam.	Size or Number	Decimal Equiv.	Outside Diam. Pitch	Root Diam.	Size or Number	Decimal Equiv.
1/16-64	0.0422	3/64	0.0469	27	0.4519	15/32	0.4687
72	0.0445	3/64	0.0469	9/16-12	0.4542	31/64	0.4844
5/64-60	0.0563	1/16	0.0625	18	0.4903	33/64	0.5156
72	0.0601	52	0.0635	27	0.5144	17/32	0.5312
3/32-48	0.0667	49	0.0730	5/8-11	0.5069	17/32	0.5312
50	0.0678	49	0.0730	12	0.5168	35/64	0.5469
7/64-48	0.0823	43	0.0890	18	0.5528	37/64	0.5781
1/8-32	0.0844	3/32	0.0937	27	0.5769	19/32	0.5937
40	0.0925	38	0.1015	11/16-11	0.5694	19/32	0.5937
9/64-40	0.1081	32	0.1160	16	0.6063	5/8	0.6250
5/32-32	0.1157	1/8	0.1250	3/4-10	0.6201	21/32	0.6562
36	0.1202	30	0.1285	12	0.6418	43/64	0.6719
11/64-32	0.1313	9/64	0.1406	16	0.6688	11/16	0.6875
3/16-24	0.1334	26	0.1470	27	0.7019	23/32	0.7187
32	0.1469	22	0.1570	13/16-10	0.6826	23/32	0.7187
13/64-24	0.1490	20	0.1610	7/8-9	0.7307	49/64	0.7656
7/32-24	0.1646	16	0.1770	12	0.7668	51/64	0.7969
32	0.1782	12	0.1890	14	0.7822	13/16	0.8125
15/64-24	0.1806	10	0.1935	18	0.8028	53/64	0.8281
1/4-20	0.1850	7	0.2010	27	0.8269	27/32	0.8437
24	0.1959	4	0.2090	15/16-9	0.7932	53/64	0.8281
27	0.2019	3	0.2130	1-8	0.8376	7/8	0.8750
28	0.2036	3	0.2130	12	0.8918	59/64	0.9219
32	0.2094	7/32	0.2187	14	0.9072	15/16	0.9375
5/16-18	0.2403	F	0.2570	27	0.9519	31/32	0.9687
20	0.2476	17/64	0.2656	1 1/8-7	0.9394	63/64	0.9844
24	0.2584	I	0.2720	12	1.0168	1 3/64	1.0469
27	0.2644	J	0.2770	1 1/4-7	1.0644	1 7/64	1.1094
32	0.2719	9/32	0.2812	12	1.1418	1 11/64	1.1719
3/8-16	0.2938	5/16	0.3125	1 3/8-6	1.1585	1 7/32	1.2187
20	0.3100	21/64	0.3281	12	1.2668	1 19/64	1.2969
24	0.3209	Q	0.3320	1 1/2-6	1.2835	1 11/32	1.3437
27	0.3269	R	0.3390	12	1.3918	1 27/64	1.4219
7/16-14	0.3447	U	0.3680	1 5/8-5 1/2	1.3888	1 29/64	1.4531
20	0.3726	25/64	0.3906	1 3/4-5	1.4902	1 9/16	1.5625
24	0.3834	X	0.3970	1 7/8-5	1.6152	1 11/16	1.6875
27	0.3894	Y	0.4040	2-4 1/2	1.7113	1 25/32	1.7812
1/2-12	0.3918	27/64	0.4219	2 1/8-4 1/2	1.8363	1 29/32	1.9062
13	0.4001	27/64	0.4219	2 1/4-4 1/2	1.9613	2 1/32	2.0312
20	0.4351	29/64	0.4531	2 3/8-4	2.0502	2 1/8	2.1250
24	0.4459	29/64	0.4531	2 1/2-4	2.1752	2 1/4	2.2500

Threads and Threading

Tap drill tables are designed to provide about 70% of effective thread. This is adequate because studies have shown that a 95% thread is only 5% stronger and risks tap breakage. There are cases where it is wise to change the thread percentage. When tapping small holes in tough material, 50% thread is sufficient. One hundred percent thread is used in non-metals and sheet metal. Use this formula to determine any thread percentage:

$$\text{Outside diameter of the thread} = \frac{0.01299 \times \textit{percentage of thread desired}}{\textit{number of threads per inch}}$$

Example: 50% thread is desired for tapping a 10–32 hole in air-hardening tool steel.

$$\text{\#10 (.190)} = \frac{0.01299 \times 50}{32} = \text{tap drill}$$

.169 (#18) = tap drill

Thread Cutting

External threads can be cut with a hand-threading die. Provide a 45° chamfer on the end of the shaft to help start the die. Threading dies can be solid, segmented with replaceable cutting edges, or with adjustable jaws. Dies have a lead angle to help get the tool started. The threading tool is fastened into a special holder. Rotate the handle of the holder clockwise while applying downward pressure on the shaft. Back the tool off every half of a turn to break the chip. Take care to start the die square to the shaft–the die will not follow the outside diameter of the shaft automatically.

Outside threads can also be cut on an engine lathe. A 60-degree threading tool is used to cut a series of grooves, starting at the same spot for each cut until the thread depth is correct. Regardless of the method used to cut external threads, the diameter of the shaft is critical. See Table 3 on pages 1817–1843.

9

Shop Recommended pages 1817, Table 3

Table 3. Standard Series and Selected Combinations — Unified Screw Threads

Nominal Size, Threads per Inch, and Series Designation[a]	Class	External[b] Allow-ance	Major Diameter Max[d]	Major Diameter Min	Major Diameter Min[e]	Pitch Diameter Max[d]	Pitch Diameter Min	UNR Minor Dia.[c] Max (Ref.)	Class	Internal[b] Minor Diameter Min	Minor Diameter Max	Pitch Diameter Min	Pitch Diameter Max	Major Diameter Min
0-80 UNF	2A	0.0005	0.0595	0.0563	—	0.0514	0.0496	0.0446	2B	0.0465	0.0514	0.0519	0.0542	0.0600
	3A	0.0000	0.0600	0.0568	—	0.0519	0.0506	0.0451	3B	0.0465	0.0514	0.0519	0.0536	0.0600
1-64 UNC	2A	0.0006	0.0724	0.0686	—	0.0623	0.0603	0.0538	2B	0.0561	0.0623	0.0629	0.0655	0.0730
	3A	0.0000	0.0730	0.0692	—	0.0629	0.0614	0.0544	3B	0.0561	0.0623	0.0629	0.0648	0.0730
1-72 UNF	2A	0.0006	0.0724	0.0689	—	0.0634	0.0615	0.0559	2B	0.0580	0.0635	0.0640	0.0665	0.0730
	3A	0.0000	0.0730	0.0695	—	0.0640	0.0626	0.0565	3B	0.0580	0.0635	0.0640	0.0659	0.0730
2-56 UNC	2A	0.0006	0.0854	0.0813	—	0.0738	0.0717	0.0642	2B	0.0667	0.0737	0.0744	0.0772	0.0860
	3A	0.0000	0.0860	0.0819	—	0.0744	0.0728	0.0648	3B	0.0667	0.0737	0.0744	0.0765	0.0860
2-64 UNF	2A	0.0006	0.0854	0.0816	—	0.0753	0.0733	0.0668	2B	0.0691	0.0753	0.0759	0.0786	0.0860
	3A	0.0000	0.0860	0.0822	—	0.0759	0.0744	0.0674	3B	0.0691	0.0753	0.0759	0.0779	0.0860
3-48 UNC	2A	0.0007	0.0983	0.0938	—	0.0848	0.0825	0.0734	2B	0.0764	0.0845	0.0855	0.0885	0.0990
	3A	0.0000	0.0990	0.0945	—	0.0855	0.0838	0.0741	3B	0.0764	0.0845	0.0855	0.0877	0.0990
3-56 UNF	2A	0.0007	0.0983	0.0942	—	0.0867	0.0845	0.0771	2B	0.0797	0.0865	0.0874	0.0902	0.0990
	3A	0.0000	0.0990	0.0949	—	0.0874	0.0858	0.0778	3B	0.0797	0.0865	0.0874	0.0895	0.0990
4-40 UNC	2A	0.0008	0.1112	0.1061	—	0.0950	0.0925	0.0814	2B	0.0849	0.0939	0.0958	0.0991	0.1120
	3A	0.0000	0.1120	0.1069	—	0.0958	0.0939	0.0822	3B	0.0849	0.0939	0.0958	0.0982	0.1120
4-48 UNF	2A	0.0007	0.1113	0.1068	—	0.0978	0.0954	0.0864	2B	0.0894	0.0968	0.0985	0.1016	0.1120
	3A	0.0000	0.1120	0.1075	—	0.0985	0.0967	0.0871	3B	0.0894	0.0968	0.0985	0.1008	0.1120
5-40 UNC	2A	0.0008	0.1242	0.1191	—	0.1080	0.1054	0.0944	2B	0.0979	0.1062	0.1088	0.1121	0.1250
	3A	0.0000	0.1250	0.1199	—	0.1088	0.1069	0.0952	3B	0.0979	0.1062	0.1088	0.1113	0.1250
5-44 UNF	2A	0.0007	0.1243	0.1195	—	0.1095	0.1070	0.0972	2B	0.1004	0.1079	0.1102	0.1134	0.1250
	3A	0.0000	0.1250	0.1202	—	0.1102	0.1083	0.0979	3B	0.1004	0.1079	0.1102	0.1126	0.1250
6-32 UNC	2A	0.0008	0.1372	0.1312	—	0.1169	0.1141	0.1000	2B	0.104	0.114	0.1177	0.1214	0.1380
	3A	0.0000	0.1380	0.1320	—	0.1177	0.1156	0.1008	3B	0.1040	0.1140	0.1177	0.1204	0.1380
6-40 UNF	2A	0.0008	0.1372	0.1321	—	0.1210	0.1184	0.1074	2B	0.111	0.119	0.1218	0.1252	0.1380
	3A	0.0000	0.1380	0.1329	—	0.1218	0.1198	0.1082	3B	0.1110	0.1186	0.1218	0.1243	0.1380
8-32 UNC	2A	0.0009	0.1631	0.1571	—	0.1428	0.1399	0.1259	2B	0.130	0.139	0.1437	0.1475	0.1640
	3A	0.0000	0.1640	0.1580	—	0.1437	0.1415	0.1268	3B	0.1300	0.1389	0.1437	0.1465	0.1640
8-36 UNF	2A	0.0008	0.1632	0.1577	—	0.1452	0.1424	0.1301	2B	0.134	0.142	0.1460	0.1496	0.1640
	3A	0.0000	0.1640	0.1585	—	0.1460	0.1439	0.1309	3B	0.1340	0.1416	0.1460	0.1487	0.1640

Size of thread desired

Minimum and maximum limits of shaft diameter before threading. Use Class 2 for general purpose threads.

9

Table 3 continues through page 1842

Threads and Threading

Index navigation paths and key words:

Threads can take on a variety of different shapes for particular applications. Transmission of motion, rifle barrels, machine shafts, and part conveyors are examples of applications that require special thread forms.

Index navigation paths and key words:

Thread grinding is used for precision applications, gages, standards, and for hardened parts. This operation is done on a *cylindrical grinder*, usually between centers. Ground threads are extremely accurate. The form of the thread is dressed on a grinding wheel and the wheel is introduced into the workpiece, either by plunging into a solid blank or following pre-machined threads.

9

SECTION 9

Q **For more information on** *cylindrical grinding,* **see Section 6, Unit I in this guide: Machining Operations**

Analyze, Evaluate, & Implement

Carefully open a 0–1 inch micrometer all the way until the thimble is unthreaded and becomes free of the frame. Observe the external ground threads on the spindle and the internal ground threads in the female component. How many threads per inch are on an English micrometer spindle?

Apply a drop of instrument oil on the spindle threads and reassemble. The micrometer can only go together one way. Advance the spindle and close the micrometer gently on a clean business card. Draw the card out and carefully close the measuring faces. Is the zero line on the sleeve coincident with the zero in the thimble as in Figure 9.3?

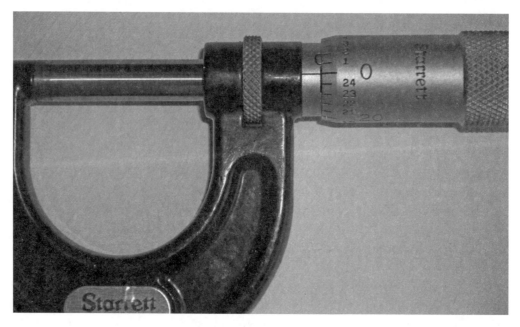

Figure 9.3 Coincident line

Threads may be cut by milling with the single-cutter method or the multiple cutter method. Internal or external threads are machined by a single cutter or a series of single cutters arranged like a row of teeth.

Threads and Threading

🔍 **Index navigation paths and key words:**

— threads and threading/thread grinding/multi-ribbed wheels/page 2054
— threads and threading/thread milling/ CNC/page 1292
— threads and threading/milling/pages 2058–2059

Unit O: Simple, Compound, Differential, and Block Indexing
pages 2079–2120

Shop Recommended page 2079, Definition

SIMPLE, COMPOUND, DIFFERENTIAL, AND BLOCK INDEXING

Milling Machine Indexing.—Positioning a workpiece at a precise angle or interval of rotation for a machining operation is called indexing. A dividing head is a milling machine attachment that provides this fine control of rotational positioning through a combination of a crank-operated worm and worm gear, and one or more indexing plates with several circles of evenly spaced holes to measure partial turns of the worm crank. The indexing crank carries a movable indexing pin that can be inserted into and withdrawn from any of the holes in a given circle with an adjustment provided for changing the circle that the indexing pin tracks.

🔍 **Index navigation paths and key words:**

— indexing/angular/pages 2086–2103
— indexing/milling machine/pages 2079–2085
— indexing/simple and differential/pages 2107–2112

9

ASSIGNMENT

List the key terms and give a definition of each.

Chamfer	Unified National Fine	Nominal Size
Crest	Major Diameter	Pitch Diameter
Thread Series	Minor Diameter	Root
Unified National Coarse		

SECTION 9

APPLY IT! PART 1

Refer to drawing # 123–456 for the following questions. (Use Table 3 on page 1817 for questions 1 and 2.)

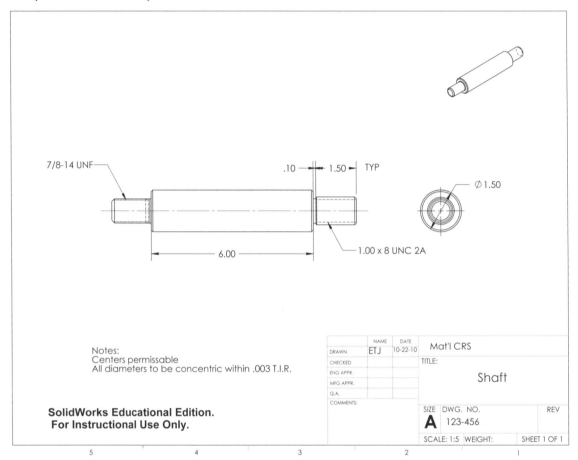

1. What are the high and low limits of size for the 1.00 diameter?

2. What are the high and low limits of size for the 7/8 diameter?

3. What is the pitch of the 1–8 UNC thread? The 7/8–14 thread?

4. What does UNC stand for?

5. How many full threads are on the 1–8 diameter? On the 7/8–14 diameter?

6. What type of material is the shaft made from?

7. One revolution of a nut on the 1–8 diameter thread will advance the nut how far?

8. What is the pitch diameter of a 7/8–14 thread?

9. Why is knowing the pitch diameter of a thread a helpful?

10. What does the designation 2A mean?

APPLY IT! PART 2

Refer to Drawing #123-457 for the following questions.

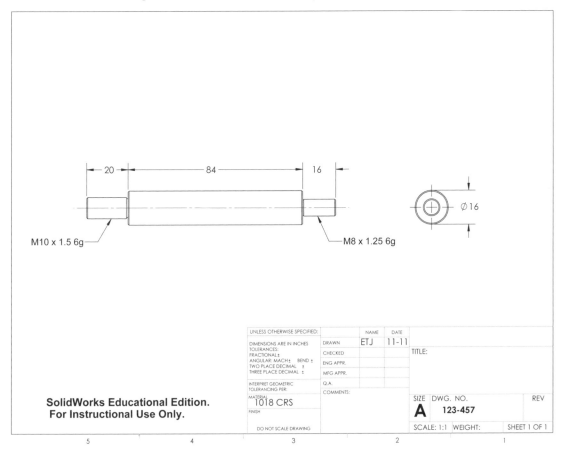

1. With regard to the thread designation M8 × 1.25–6g, what does M8 mean? 1.25? 6g?

2. What is the thread angle of the threads shown on this drawing?

3. What is the major diameter of the M8 thread? (nominal size) The M10 thread? (nominal size)

4. If a machined finish is required all over, what size bar stock (in inches) would be used for this shaft?

5. How many complete threads are on the M10 diameter?

9

SECTION 9

APPLY IT! PART 3 TAP DRILL QUESTIONS

1. What size tap drill is used to tap a 3/8–16 thread in plexiglass for 100% thread?

2. What size tap drill is used to tap a #6–32 thread in 4140 heat treated steel for 50% thread?

3. For maximum holding power, how deep should a 1/2–13 hole be tapped? A 5/16–18?

9

Section 10

GEARS, SPLINES, and CAMS

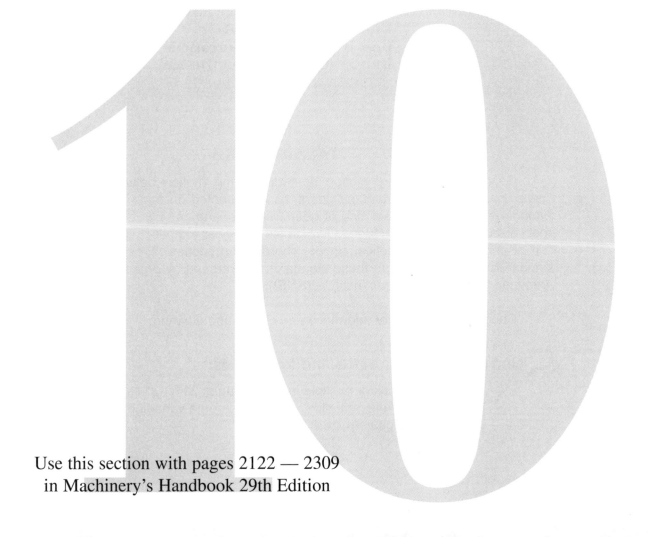

Use this section with pages 2122 — 2309
in Machinery's Handbook 29th Edition

SECTION 10

Navigation Overview

 All Units in this Section covered with: Navigation Assistant

The Navigation Assistant helps find information in the MH29 Primary Index. The Primary Index is located in the back of the book on pages 2701–2788 and is set up alphabetically by subject. Watch for the magnifying glass throughout the section for navigation hints.

May I help you?

Unit A	Gears and Gearing
Unit B	Hypoid and Bevel Gearing
Unit C	Worm Gearing
Unit D	Helical Gearing
Unit E	Other Gear Types
Unit F	Checking Gear Sizes
Unit G	Gear Materials
Unit H	Spines and Serrations
Unit I	Cams and Cam Design

Introduction

Gears transmit rotary motion or reciprocating motion between machine parts. There are many styles of toothed gears, as represented on page 2125 of *Machinery's Handbook 29*. The manufacturing of gears, splines, and cams is a highly technical and specialized area of manufacturing. The profile of a gear tooth is a complicated combination of angles, radii, theoretical circles, chords, and pitches. These characteristics are standardized. Compliance to these standards is overseen by organizations such as the American National Standards Institute (ANSI).

Check these pages for information on Gears and Gearing:

 Index navigation paths and key words:

— gear cutting/block or multiple indexing/pages 2117–2118
— gear materials/effect of alloying metals/chrome vanadium/page 2243
— pitch/circle, of gears/page 2130

10

— pitch/gear/page 2131
— gears and gearing/backlash/pages 2163—2169; 2235
— gears and gearing/checking/gear sizes/pages 2221–2239

Index navigation paths and key words:

Check these pages for information on Splines and Cams:
— splines/page 2255
— splines/data and reference dimensions/page 2265
— cams and cam design/pages 1174–1175; 2284–2309

10

Section 11

MACHINE ELEMENTS

Use this section with pages 2310 — 2650
in Machinery's Handbook 29th Edition

SECTION 11

Key Terms

Keyseat	Wire EDM
Keyway	Keyseat Alignment
Woodruff Key	Tolerance
Broach	Clearance Fit
Keyseating machine	Interference Fit
Broach	

11

210

Learning Objectives:

After studying this unit you should be able to:

- **Select the proper size key for a given shaft diameter.**
- **Interpret the size of a Woodruff key by its part number.**
- **State the three methods of machining a keyway.**
- **Calculate allowable tolerances for machining keyways and keyseats.**
- **State the difference between a clearance fit and an interference fit.**

Unit E: **Keys and Keyseats pages 2460–2483**

A key is a small component of an assembly that is used to align shafts, hubs, pulleys, handles, cutters, and gears. Metal keys lie in a shallow groove with part of the key in each component. They are used to transmit motion through shafts by eliminating slippage. Keys are square, rectangular (*flat*), or semi-circular (*woodruff*). Like electrical fuses, keys are designed to fail, or shear, before more expensive components are damaged.

In an assembly with a shaft and a hub, the *keyway* is the groove in the hub and the *keyseat* is the groove in the shaft (see Figure 11.1). The fits between the components of a keyed assembly is closely controlled.

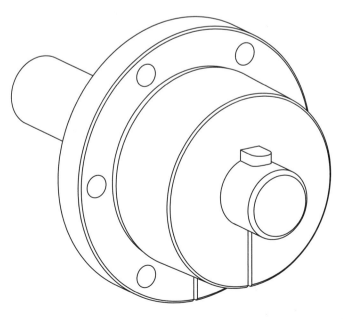

Figure 11.1 Keyed assembly of shaft and hub

11

SECTION 11

Woodruff Keys

Index navigation path and key words:

— keys and keyseats/woodruff keys and keyseats/page 2477

Woodruff keys are semi-circular and fit into a semi-circular groove that is shallower than the key leaving part of the key extended in the mating part. Woodruff keys are produced by cutters of the same name and are identified by a four or five digit number. The last two numbers give the diameter of the circle in eighths of an inch that would contain the key if it were a complete circle. The numbers preceding the last two numbers are the numbers of 1/32ths of an inch in the keys' width (or thickness).

Design Considerations

For the machinist, keys and keyways are produced from an engineering drawing, with dimensions and tolerances already calculate by the designer, but the machinist should be aware of some basic practices regarding keys and keyways:

- For square and rectangular keys, key length should be less than 10 times the key width.
- There is a uniform relationship between the shaft size and the key size.
- Woodruff keys, square keys, and rectangular keys are selected based on the type of service and the shaft diameter.

Machining Methods

Keyways (in a hub, not in the shaft) are produced by one of the following methods:

- Keyseating machine
- Wire EDM (electrical discharge machine)
- Broach

The **keyseating machine** is actually misnamed because it cuts *keyways* in hubs, gears, and pulleys by drawing or pushing a *broach* through a finished bore (hole) (see Figure 11.2).

11

212

Machine Elements

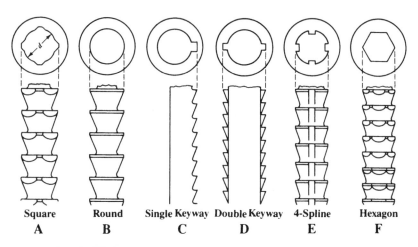

Square	Round	Single Keyway	Double Keyway	4-Spline	Hexagon
A	B	C	D	E	F

Figure 11.2 Broaches used in a keyseating machine

Wire EDM

 Index navigation path and key words:

— wire/EDM/pages 1393, 1402

Wire EDM *(electrical discharge machine)* works like a tiny band saw blade except it uses a wire (.002 to .010) that follows a programmed path. The wire does not actually touch the electrically conductive workpiece. Rapid electrical impulses vaporize the material. Wire EDM is very accurate, but the process is slow. It is typical for a Wire EDM machine to run for long periods of time unattended.

Broaching

Index navigation path and key words:

— broaching/pages 979–985

Broaching can be done in machines other than a keyseating machine. A broach can be pushed through a finished bore on an arbor press or setup in a vertical milling machine. A headed bushing that accepts the shape of the broach is inserted into the hole. A series of cuts are taken by adding shims to the back of the broach until the finish size is achieved.

In an emergency, a keyway cutter can be made from a high speed steel cutting tool. The machinist grinds the shape of one broach tooth on one end of an old end mill. The cutter is held securely in the spindle of a vertical milling machine and light cuts are taken by advancing the table and pulling down on the quill feed lever. Put the machine in the lowest speed possible, shut off the spindle, line up the tool perpendicular to the cut and apply the brake (if there is one). When the machine is in a low speed, it prevents the spindle from turning easily.

11

SECTION 11

Shop Recommended page 2472, Table 1

Table 1. Key Size Versus Shaft Diameter *ANSI B17.1-1967 (R2008)*

Nominal Shaft Diameter		Nominal Key Size			Normal Keyseat Depth	
			Height, *H*		*H/2*	
Over	To (Incl.)	Width, *W*	Square	Rectangular	Square	Rectangular
$\frac{5}{16}$	$\frac{7}{16}$	$\frac{3}{32}$	$\frac{3}{32}$	...	$\frac{3}{64}$	...
$\frac{7}{16}$	$\frac{9}{16}$	$\frac{1}{8}$	$\frac{1}{8}$	$\frac{3}{32}$	$\frac{1}{16}$	$\frac{3}{64}$
$\frac{9}{16}$	$\frac{7}{8}$	$\frac{3}{16}$	$\frac{3}{16}$	$\frac{1}{8}$	$\frac{3}{32}$	$\frac{1}{16}$
$\frac{7}{8}$	$1\frac{1}{4}$	$\frac{1}{4}$	$\frac{1}{4}$	$\frac{3}{16}$	$\frac{1}{8}$	$\frac{3}{32}$
$1\frac{1}{4}$	$1\frac{3}{8}$	$\frac{5}{16}$	$\frac{5}{16}$	$\frac{1}{4}$	$\frac{5}{32}$	$\frac{1}{8}$
$1\frac{3}{8}$	$1\frac{3}{4}$	$\frac{3}{8}$	$\frac{3}{8}$	$\frac{1}{4}$	$\frac{3}{16}$	$\frac{1}{8}$
$1\frac{3}{4}$	$2\frac{1}{4}$	$\frac{1}{2}$	$\frac{1}{2}$	$\frac{3}{8}$	$\frac{1}{4}$	$\frac{3}{16}$
$2\frac{1}{4}$	$2\frac{3}{4}$	$\frac{5}{8}$	$\frac{5}{8}$	$\frac{7}{16}$	$\frac{5}{16}$	$\frac{7}{32}$
$2\frac{3}{4}$	$3\frac{1}{4}$	$\frac{3}{4}$	$\frac{3}{4}$	$\frac{1}{2}$	$\frac{3}{8}$	$\frac{1}{4}$
$3\frac{1}{4}$	$3\frac{3}{4}$	$\frac{7}{8}$	$\frac{7}{8}$	$\frac{5}{8}$	$\frac{7}{16}$	$\frac{5}{16}$
$3\frac{3}{4}$	$4\frac{1}{2}$	1	1	$\frac{3}{4}$	$\frac{1}{2}$	$\frac{3}{8}$
$4\frac{1}{2}$	$5\frac{1}{2}$	$1\frac{1}{4}$	$1\frac{1}{4}$	$\frac{7}{8}$	$\frac{5}{8}$	$\frac{7}{16}$
$5\frac{1}{2}$	$6\frac{1}{2}$	$1\frac{1}{2}$	$1\frac{1}{2}$	1	$\frac{3}{4}$	$\frac{1}{2}$
Square Keys preferred for shaft diameters above this line; rectangular keys, below						
$6\frac{1}{2}$	$7\frac{1}{2}$	$1\frac{3}{4}$	$1\frac{3}{4}$	$1\frac{1}{2}$a	$\frac{7}{8}$	$\frac{3}{4}$
$7\frac{1}{2}$	9	2	2	$1\frac{1}{2}$	1	$\frac{3}{4}$
9	11	$2\frac{1}{2}$	$2\frac{1}{2}$	$1\frac{3}{4}$	$1\frac{1}{4}$	$\frac{7}{8}$

Nominal size is the size used for the general identification of the shaft size.

Width of key recommended for the shaft shown on the left.

Square keys have the same *width W* and *height H*. Rectangular keys are sometimes called "flat keys" and are recommended for larger shaft diameters.

This is the nominal depth of the *keyseat*. The actual machined depth is shown in MHB 29, page 2473; Table 2

The *Keyseat Alignment Tolerance* is the amount of acceptable error between the centerline of the key seat and the centerline of the shaft or bore (see Figure 11.3). The tolerance is .002 for keyseats up to and including 4″ and .0005 per inch for keyseats from 4″ to 10″ in length. When machining keyways and keyseats, take care to precisely locate the centerline of the shaft or bore with a dial indicator.

11

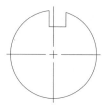

Figure 11.3 This keyseat is obviously not centered on the shaft.

Shop Recommended page 2473, Table 2

Dimension "S" is critical. For small keyseats, use a micrometer to measure over a gage pin or dowel placed in the bottom of the keyseat. Don't forget to subtract the diameter of the pin for the correct reading.

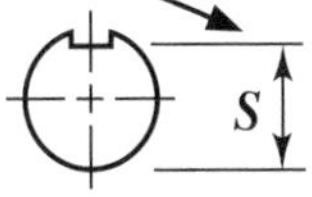

Dimension "T" can be carefully measured with a dial caliper.

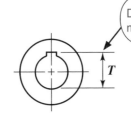

Table 2. Depth Control Values S and T for Shaft and Hub
ANSI B17.1-1967 (R2008)

On tapered hubs, the keyseat is parallel to the centerline of the shaft and a tapered key is used.

Nominal Shaft Diameter	Shafts, Parallel and Taper		Hubs, Parallel		Hubs, Taper	
	Square	Rectangular	Square	Rectangular	Square	Rectangular
	S	S	T	T	T	T
½	0.430	0.445	0.560	0.544	0.535	0.519
⁹⁄₁₆	0.493	0.509	0.623	0.607	0.598	0.582
⅝	0.517	0.548	0.709	0.678	0.684	0.653
¹¹⁄₁₆	0.581	0.612	0.773	0.742	0.748	0.717
¾	0.644	0.676	0.837	0.806	0.812	0.781
¹³⁄₁₆	0.708	0.739	0.900	0.869	0.875	0.844
⅞	0.771	0.802	0.964	0.932	0.939	0.907
¹⁵⁄₁₆	0.796	0.827	1.051	1.019	1.026	0.994
1	0.859	0.890	1.114	1.083	1.089	1.058
1¹⁄₁₆	0.923	0.954	1.178	1.146	1.153	1.121
1⅛	0.986	1.017	1.241	1.210	1.216	1.185
1³⁄₁₆	1.049	1.080	1.304	1.273	1.279	1.248
1¼	1.112	1.144	1.367	1.336	1.342	1.311
1⁵⁄₁₆	1.137	1.169	1.455	1.424	1.430	1.399
1⅜	1.201	1.232	1.518	1.487	1.493	1.462
1⁷⁄₁₆	1.225	1.288	1.605	1.543	1.580	1.518
1½	1.289	1.351	1.669	1.606	1.644	1.581
1⁹⁄₁₆	1.352	1.415	1.732	1.670	1.707	1.645
1⅝	1.416	1.478	1.796	1.733	1.771	1.708
1¹¹⁄₁₆	1.479	1.541	1.859	1.796	1.834	1.771
1¾	1.542	1.605	1.922	1.860	1.897	1.835
1¹³⁄₁₆	1.527	1.590	2.032	1.970	2.007	1.945
1⅞	1.591	1.654	2.096	2.034	2.071	2.009
1¹⁵⁄₁₆	1.655	1.717	2.160	2.097	2.135	2.072
2	1.718	1.781	2.223	2.161	2.198	2.136
2¹⁄₁₆	1.782	1.844	2.287	2.224	2.262	2.199
2⅛	1.845	1.908	2.350	2.288	2.325	2.263
2³⁄₁₆	1.909	1.971	2.414	2.351	2.389	2.326
2¼	1.972	2.034	2.477	2.414	2.452	2.389
2⁵⁄₁₆	1.957	2.051	2.587	2.493	2.562	2.468
2⅜	2.021	2.114	2.651	2.557	2.626	2.532
2⁷⁄₁₆	2.084	2.178	2.714	2.621	2.689	2.596
2½	2.148	2.242	2.778	2.684	2.753	2.659

11

215

SECTION 11

Shop Recommended page 2475, Table 3

Table 3. ANSI Standard Plain and Gib Head Keys *ANSI B17.1-1967 (R2008)*

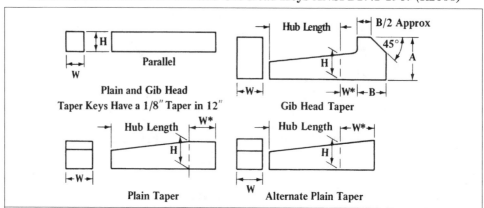

The keys shown in Table 3 on page 2475 in *Machinery's Handbook 29* are used in tapered shafts and bores. A gib head taper key can be driven out of the shaft-hub assembly without further disassembly (see Figure 11.4).

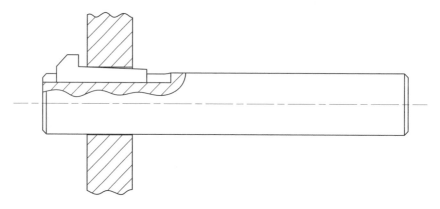

Figure 11.4 Gib head taper key in assembly

Shop Recommended page 2476, Table 4

Table 4 gives the tolerances for the machined width and depth of the keyseat. The class of fit is determined by the application. A *clearance fit (CL)* is when there is a designed allowance between mating components. An *interference fit (INT)* is when one part is designed to be physically larger than the mating part.

Table 4. ANSI Standard Fits for Parallel and Taper Keys *ANSI B17.1-1967 (R2008)*

Type of Key	Key Width Over	Key Width To (Incl.)	Side Fit Width Tolerance Key	Side Fit Width Tolerance Key-Seat	Side Fit Fit Range[a]	Top and Bottom Fit Depth Tolerance Key	Top and Bottom Fit Depth Tolerance Shaft Key-Seat	Top and Bottom Fit Depth Tolerance Hub Key-Seat	Top and Bottom Fit Fit Range[a]
Class 1 Fit for Parallel Keys								
Square	…	½	+0.000 / −0.002	+0.002 / −0.000	0.004 CL / 0.000	+0.000 / −0.002	+0.000 / −0.015	+0.010 / −0.000	0.032 CL / 0.005 CL
	½	¾	+0.000 / −0.002	+0.003 / −0.000	0.005 CL / 0.000	+0.000 / −0.002	+0.000 / −0.015	+0.010 / −0.000	0.032 CL / 0.005 CL
	¾	1	+0.000 / −0.003	+0.003 / −0.000	0.006 CL / 0.000	+0.000 / −0.003	+0.000 / −0.015	+0.010 / −0.000	0.033 CL / 0.005 CL
	1	1½	+0.000 / −0.003	+0.004 / −0.000	0.007 CL / 0.000	+0.000 / −0.003	+0.000 / −0.015	+0.010 / −0.000	0.033 CL / 0.005 CL
	1½	2½	+0.000 / −0.004	+0.004 / −0.000	0.008 CL / 0.000	+0.000 / −0.004	+0.000 / −0.015	+0.010 / −0.000	0.034 CL / 0.005 CL
	2½	3½	+0.000 / −0.006	+0.004 / −0.000	0.010 CL / 0.000	+0.000 / −0.006	+0.000 / −0.015	+0.010 / −0.000	0.036 CL / 0.005 CL
Rectangular	…	½	+0.000 / −0.003	+0.002 / −0.000	0.005 CL / 0.000	+0.000 / −0.003	+0.000 / −0.015	+0.010 / −0.000	0.033 CL / 0.005 CL
	½	¾	+0.000 / −0.003	+0.003 / −0.000	0.006 CL / 0.000	+0.000 / −0.003	+0.000 / −0.015	+0.010 / −0.000	0.033 CL / 0.005 CL
	¾	1	+0.000 / −0.004	+0.003 / −0.000	0.007 CL / 0.000	+0.000 / −0.004	+0.000 / −0.015	+0.010 / −0.000	0.034 CL / 0.005 CL
	1	1½	+0.000 / −0.004	+0.004 / −0.000	0.008 CL / 0.000	+0.000 / −0.004	+0.000 / −0.015	+0.010 / −0.000	0.034 CL / 0.005 CL
	1½	3	+0.000 / −0.005	+0.004 / −0.000	0.009 CL / 0.000	+0.000 / −0.005	+0.000 / −0.015	+0.010 / −0.000	0.035 CL / 0.005 CL
	3	4	+0.000 / −0.006	+0.004 / −0.000	0.010 CL / 0.000	+0.000 / −0.006	+0.000 / −0.015	+0.010 / −0.000	0.036 CL / 0.005 CL
	4	6	+0.000 / −0.008	+0.004 / −0.000	0.012 CL / 0.000	+0.000 / −0.008	+0.000 / −0.015	+0.010 / −0.000	0.038 CL / 0.005 CL
	6	7	+0.000 / −0.013	+0.004 / −0.000	0.017 CL / 0.000	+0.000 / −0.013	+0.000 / −0.015	+0.010 / −0.000	0.043 CL / 0.005 CL
Class 2 Fit for Parallel and Taper Keys								
Parallel Square	…	1¼	+0.001 / −0.000	+0.002 / −0.000	0.002 CL / 0.001 INT	+0.001 / −0.000	+0.000 / −0.015	+0.010 / −0.000	0.030 CL / 0.004 CL
	1¼	3	+0.002 / −0.000	+0.002 / −0.000	0.002 CL / 0.002 INT	+0.002 / −0.000	+0.000 / −0.015	+0.010 / −0.000	0.030 CL / 0.003 CL
	3	3½	+0.003 / −0.000	+0.002 / −0.000	0.002 CL / 0.003 INT	+0.003 / −0.000	+0.000 / −0.015	+0.010 / −0.000	0.030 CL / 0.002 CL
Parallel Rectangular	…	1¼	+0.001 / −0.000	+0.002 / −0.000	0.002 CL / 0.001 INT	+0.005 / −0.005	+0.000 / −0.015	+0.010 / −0.000	0.035 CL / 0.000 CL
	1¼	3	+0.002 / −0.000	+0.002 / −0.000	0.002 CL / 0.002 INT	+0.005 / −0.005	+0.000 / −0.015	+0.010 / −0.00	0.035 CL / 0.000 CL
	3	7	+0.003 / −0.000	+0.002 / −0.000	0.002 CL / 0.003 INT	+0.005 / −0.005	+0.000 / −0.015	+0.010 / −0.000	0.035 CL / 0.000 CL
Taper	…	1¼	+0.001 / −0.000	+0.002 / −0.000	0.002 CL / 0.001 INT	+0.005 / −0.000	+0.000 / −0.015	+0.010 / −0.000	0.005 CL / 0.025 INT
	1¼	3	+0.002 / −0.000	+0.002 / −0.000	0.002 CL / 0.002 INT	+0.005 / −0.000	+0.000 / −0.015	+0.010 / −0.000	0.005 CL / 0.025 INT
	3	b	+0.003 / −0.000	+0.002 / −0.000	0.002 CL / 0.003 INT	+0.005 / −0.000	+0.000 / −0.015	+0.010 / −0.000	0.005 CL / 0.025 INT

[a] Limits of variation. CL = Clearance; INT = Interference.

[b] To (Incl.) 3½-inch Square and 7-inch Rectangular key widths.

All dimensions are given in inches. See also text on page 2460.

Side-note callouts:

The side fit is the width of the key and keyseat and is more critical than the depth.

For key-keyseat assemblies 1/2″ and smaller, a width clearance of 0.000 to 0.004 is required for a Class 1 fit.

For a 1″ rectangular key, a height tolerance of 0.034 to 0.005 is required for a class 1 fit.

Class 2 fits are for general purpose applications.

11

SECTION 11

🔍 **Navigation Hint: For more information on clearance and inter-ference fits, see Section 4 Unit B, Dimensioning, Gaging, and Measuring, MH29 pages 609–750**

Shop Recommended page 2477, Table 7

Set screws are sometimes used over the key for strength (see Figure 11.5). The female component is threaded to accept the set screw. The size of the set screw is based on the shaft diameter and key width. For example, a 1″ shaft having a 1/4″ key width should use a 5/16 diameter set screw. For extra security against loosening, install a second set screw in the same hole behind the first screw.

Table 7. Set Screws for Use Over Keys *ANSI B17.1-1967 (R2008)*

Nom. Shaft Dia. Over	Nom. Shaft Dia. To (Incl.)	Nom. Key Width	Set Screw Dia.	Nom. Shaft Dia. Over	Nom. Shaft Dia. To (Incl.)	Nom. Key Width	Set Screw Dia.
$\frac{5}{16}$	$\frac{7}{16}$	$\frac{3}{32}$	No. 10	$2\frac{1}{4}$	$2\frac{3}{4}$	$\frac{5}{8}$	$\frac{1}{2}$
$\frac{7}{16}$	$\frac{9}{16}$	$\frac{1}{8}$	No. 10	$2\frac{3}{4}$	$3\frac{1}{4}$	$\frac{3}{4}$	$\frac{5}{8}$
$\frac{9}{16}$	$\frac{7}{8}$	$\frac{3}{16}$	$\frac{1}{4}$	$3\frac{1}{4}$	$3\frac{3}{4}$	$\frac{7}{8}$	$\frac{3}{4}$
$\frac{7}{8}$	$1\frac{1}{4}$	$\frac{1}{4}$	$\frac{5}{16}$	$3\frac{3}{4}$	$4\frac{1}{2}$	1	$\frac{3}{4}$
$1\frac{1}{4}$	$1\frac{3}{8}$	$\frac{5}{16}$	$\frac{3}{8}$	$4\frac{1}{2}$	$5\frac{1}{2}$	$1\frac{1}{4}$	$\frac{7}{8}$
$1\frac{3}{8}$	$1\frac{3}{4}$	$\frac{3}{8}$	$\frac{3}{8}$	$5\frac{1}{2}$	$6\frac{1}{2}$	$1\frac{1}{2}$	1
$1\frac{3}{4}$	$2\frac{1}{4}$	$\frac{1}{2}$	$\frac{1}{2}$	…	…	…	…

All dimensions are given in inches.

These set screw diameter selections are offered as a guide but their use should be dependent upon design considerations.

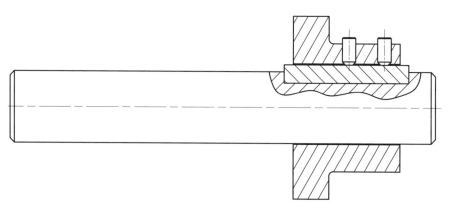

Figure 11.5 Section view of shaft and hub with key secured with set screws

218

Machine Elements

Shop Recommended page 2480, Table 10

Woodruff keys fit into a semi-circular groove that is shallower than the key leaving part of the key extended in the mating part (see Figure 11.6). The last two numbers of the key number are the diameter of the circle in eighths of an inch that would contain the key if it were a complete circle. The numbers preceding the last two numbers are the numbers of 1/32ths of an inch in the keys' width (or thickness).

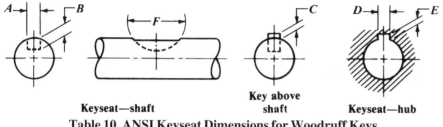

Keyseat—shaft   **Key above shaft**   **Keyseat—hub**

The reason dimension lines *B*, *C*, and *E* are on an angle is to make it clear that the dimension is taken from the edge of the key, not from the top of the diameter of the hole or shaft.

Table 10. ANSI Keyseat Dimensions for Woodruff Keys
ANSI B17.2-1967 (R2008)

Key No.	Nominal Size Key	Keyseat—Shaft					Key Above Shaft	Keyseat—Hub	
		Width A^a		Depth B	Diameter F		Height C	Width D	Depth E
		Min.	Max.	+0.005 −0.000	Min.	Max.	+0.005 −0.005	+0.002 −0.000	+0.005 −0.000
202	$\frac{1}{16} \times \frac{1}{4}$	0.0615	0.0630	0.0728	0.250	0.268	0.0312	0.0635	0.0372
202.5	$\frac{1}{16} \times \frac{5}{16}$	0.0615	0.0630	0.1038	0.312	0.330	0.0312	0.0635	0.0372
302.5	$\frac{3}{32} \times \frac{5}{16}$	0.0928	0.0943	0.0882	0.312	0.330	0.0469	0.0948	0.0529
203	$\frac{1}{16} \times \frac{3}{8}$	0.0615	0.0630	0.1358	0.375	0.393	0.0312	0.0635	0.0372
303	$\frac{3}{32} \times \frac{3}{8}$	0.0928	0.0943	0.1202	0.375	0.393	0.0469	0.0948	0.0529
403	$\frac{1}{8} \times \frac{3}{8}$	0.1240	0.1255	0.1045	0.375	0.393	0.0625	0.1260	0.0685
204	$\frac{1}{16} \times \frac{1}{2}$	0.0615	0.0630	0.1668	0.500	0.518	0.0312	0.0635	0.0372
304	$\frac{3}{32} \times \frac{1}{2}$	0.0928	0.0943	0.1511	0.500	0.518	0.0469	0.0948	0.0529
404	$\frac{1}{8} \times \frac{1}{2}$	0.1240	0.1255	0.1355	0.500	0.518	0.0625	0.1260	0.0685
305	$\frac{3}{32} \times \frac{5}{8}$	0.0928	0.0943	0.1981	0.625	0.643	0.0469	0.0948	0.0529
405	$\frac{1}{8} \times \frac{5}{8}$	0.1240	0.1255	0.1825	0.625	0.643	0.0625	0.1260	0.0685
505	$\frac{5}{32} \times \frac{5}{8}$	0.1553	0.1568	0.1669	0.625	0.643	0.0781	0.1573	0.0841
605	$\frac{3}{16} \times \frac{5}{8}$	0.1863	0.1880	0.1513	0.625	0.643	0.0937	0.1885	0.0997
406	$\frac{1}{8} \times \frac{3}{4}$	0.1240	0.1255	0.2455	0.750	0.768	0.0625	0.1260	0.0685
506	$\frac{5}{32} \times \frac{3}{4}$	0.1553	0.1568	0.2299	0.750	0.768	0.0781	0.1573	0.0841
606	$\frac{3}{16} \times \frac{3}{4}$	0.1863	0.1880	0.2143	0.750	0.768	0.0937	0.1885	0.0997
806	$\frac{1}{4} \times \frac{3}{4}$	0.2487	0.2505	0.1830	0.750	0.768	0.1250	0.2510	0.1310
507	$\frac{5}{32} \times \frac{7}{8}$	0.1553	0.1568	0.2919	0.875	0.895	0.0781	0.1573	0.0841
607	$\frac{3}{16} \times \frac{7}{8}$	0.1863	0.1880	0.2763	0.875	0.895	0.0937	0.1885	0.0997
707	$\frac{7}{32} \times \frac{7}{8}$	0.2175	0.2193	0.2607	0.875	0.895	0.1093	0.2198	0.1153
807	$\frac{1}{4} \times \frac{7}{8}$	0.2487	0.2505	0.2450	0.875	0.895	0.1250	0.2510	0.1310
608	$\frac{3}{16} \times 1$	0.1863	0.1880	0.3393	1.000	1.020	0.0937	0.1885	0.0997
708	$\frac{7}{32} \times 1$	0.2175	0.2193	0.3237	1.000	1.020	0.1093	0.2198	0.1153
808	$\frac{1}{4} \times 1$	0.2487	0.2505	0.3080	1.000	1.020	0.1250	0.2510	0.1310
1008	$\frac{5}{16} \times 1$	0.3111	0.3130	0.2768	1.000	1.020	0.1562	0.3135	0.1622
1208	$\frac{3}{8} \times 1$	0.3735	0.3755	0.2455	1.000	1.020	0.1875	0.3760	0.1935
609	$\frac{3}{16} \times 1\frac{1}{8}$	0.1863	0.1880	0.3853	1.125	1.145	0.0937	0.1885	0.0997
709	$\frac{7}{32} \times 1\frac{1}{8}$	0.2175	0.2193	0.3697	1.125	1.145	0.1093	0.2198	0.1153
809	$\frac{1}{4} \times 1\frac{1}{8}$	0.2487	0.2505	0.3540	1.125	1.145	0.1250	0.2510	0.1310

For a #406 key, the "6" is the number of eighths of an inch in keys circle. Six eighths = 3/4 diameter. Four is the number of 32nds in the keys thickness. Four 32nds = 1/8.

continued on next page

219

11

809	$\frac{1}{4} \times 1\frac{1}{8}$	0.2487	0.2505	0.3540	1.125	1.145	0.1250	0.2510	0.1310
1009	$\frac{5}{16} \times 1\frac{1}{8}$	0.3111	0.3130	0.3228	1.125	1.145	0.1562	0.3135	0.1622
610	$\frac{3}{16} \times 1\frac{1}{4}$	0.1863	0.1880	0.4483	1.250	1.273	0.0937	0.1885	0.0997
710	$\frac{7}{32} \times 1\frac{1}{4}$	0.2175	0.2193	0.4327	1.250	1.273	0.1093	0.2198	0.1153
810	$\frac{1}{4} \times 1\frac{1}{4}$	0.2487	0.2505	0.4170	1.250	1.273	0.1250	0.2510	0.1310
1010	$\frac{5}{16} \times 1\frac{1}{4}$	0.3111	0.3130	0.3858	1.250	1.273	0.1562	0.3135	0.1622
1210	$\frac{3}{8} \times 1\frac{1}{4}$	0.3735	0.3755	0.3545	1.250	1.273	0.1875	0.3760	0.1935
811	$\frac{1}{4} \times 1\frac{3}{8}$	0.2487	0.2505	0.4640	1.375	1.398	0.1250	0.2510	0.1310
1011	$\frac{5}{16} \times 1\frac{3}{8}$	0.3111	0.3130	0.4328	1.375	1.398	0.1562	0.3135	0.1622

> The last three fields are important for the machining of the keyseat: Height of the installed key, Width, and Depth.

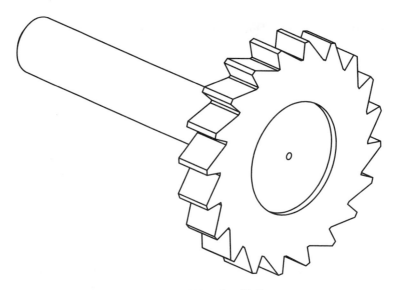

Figure 11.6 Woodruff Cutter

Shop Recommended page 2483, Table 11

Table 11 on page 2483 of *Machinery's Handbook 29* gives the dimension *M*, which is helpful. When machining a keyseat, the outside diameter of the shaft is an easily referenced surface. This surface becomes a beginning point or *0* on the machine dial or digital readout. This value is added to dimension *D* to calculate the total depth of the cutter from the surface of the shaft. The best way to measure the keyseat is to insert the key and measure dimension *J* because this is the way the assembly is going to be used.

11

Table 11. Finding Depth of Keyseat and Distance from Top of Key to Bottom of Shaft

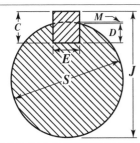

For milling keyseats, the total depth to feed cutter in from outside of shaft to bottom of keyseat is $M + D$, where D is depth of keyseat.

For checking an assembled key and shaft, caliper measurement J between top of key and bottom of shaft is used.

$$J = S - (M + D) + C$$

where C is depth of key. For Woodruff keys, dimensions C and D can be found in Tables 8 through 10. Assuming shaft diameter S is normal size, the tolerance on dimension J for Woodruff keys in keyslots are + 0.000, −0.010 inch.

Dia. of Shaft, S Inches	Width of Keyseat, E														
	$\frac{1}{16}$	$\frac{3}{32}$	$\frac{1}{8}$	$\frac{5}{32}$	$\frac{3}{16}$	$\frac{7}{32}$	$\frac{1}{4}$	$\frac{5}{16}$	$\frac{3}{8}$	$\frac{7}{16}$	$\frac{1}{2}$	$\frac{9}{16}$	$\frac{5}{8}$	$\frac{11}{16}$	$\frac{3}{4}$
	Dimension M, Inch														
0.3125	.0032	...	...	...	...	...	...	...	...	...	...	...	...	...	...
0.3437	.0029	.0065	...	...	...	...	...	...	...	...	...	...	...	...	...
0.3750	.0026	.0060	.0107	...	...	...	...	...	...	...	...	...	...	...	...
0.4060	.0024	.0055	.0099	...	...	...	...	...	...	...	...	...	...	...	...
0.4375	.0022	.0051	.0091	...	...	...	...	...	...	...	...	...	...	...	...
0.4687	.0021	.0047	.0085	.0134	...	...	...	...	...	...	...	...	...	...	...
0.5000	.0020	.0044	.0079	.0125	...	...	...	...	...	...	...	...	...	...	...
0.5625	...	.0039	.0070	.0111	.0161	...	...	...	...	...	...	...	...	...	...
0.6250	...	.0035	.0063	.0099	.0144	.0198	...	...	...	...	...	...	...	...	...
0.6875	...	.0032	.0057	.0090	.0130	.0179	.0235	...	...	...	...	...	...	...	...
0.7500	...	.0029	.0052	.0082	.0119	.0163	.0214	.0341	...	...	...	...	...	...	...
0.8125	...	.0027	.0048	.0076	.0110	.0150	.0197	.0312	...	...	...	...	...	...	...
0.8750	...	.0025	.0045	.0070	.0102	.0139	.0182	.0288	...	...	...	...	...	...	...
0.9375	...	...	.0042	.0066	.0095	.0129	.0170	.0263	.0391	...	...	...	...	...	...
1.0000	...	...	.0039	.0061	.0089	.0121	.0159	.0250	.0365	...	...	...	...	...	...
1.0625	...	...	.0037	.0058	.0083	.0114	.0149	.0235	.0342	...	...	...	...	...	...
1.1250	...	...	.0035	.0055	.0079	.0107	.0141	.0221	.0322	.0443	...	...	...	...	...
1.1875	...	...	.0033	.0052	.0074	.0102	.0133	.0209	.0304	.0418	...	...	...	...	...
1.2500	...	...	.0031	.0049	.0071	.0097	.0126	.0198	.0288	.0395	...	...	...	...	...
1.3750	...	...	...	.0045	.0064	.0088	.0115	.0180	.0261	.0357	.0471	...	...	...	...
1.5000	...	...	...	.0041	.0059	.0080	.0105	.0165	.0238	.0326	.0429	...	...	...	...
1.6250	...	...	...	.0038	.0054	.0074	.0097	.0152	.0219	.0300	.0394	.0502	...	...	...
1.7500	...	...	...	...	.0050	.0069	.0090	.0141	.0203	.0278	.0365	.0464	...	...	...
1.8750	...	...	...	...	.0047	.0064	.0084	.0131	.0189	.0259	.0340	.0432	.0536	...	...
2.0000	...	...	...	...	.0044	.0060	.0078	.0123	.0177	.0242	.0318	.0404	.0501	...	...
2.1250	...	...	...	...	...	.0056	.0074	.0116	.0167	.0228	.0298	.0379	.0470	.0572	.0684
2.2500	...	...	...	...	...	...	.0070	.0109	.0157	.0215	.0281	.0357	.0443	.0538	.0643
2.3750	...	...	...	...	...	...	.0103	.0149	.0203	.0266	.0338	.0419	.0509	.0608	
2.5000	...	...	...	...	...	...	...	.0141	.0193	.0253	.0321	.0397	.0482	.0576	
2.6250	...	...	...	...	...	...	...	.0135	.0184	.0240	.0305	.0377	.0457	.0547	
2.7500	...	...	...	...	...	...	...	...	.0175	.0229	.0291	.0360	.0437	.0521	
2.8750	...	...	...	...	...	...	...	...	.0168	.0219	.0278	.0344	.0417	.0498	
3.0000	...	...	...	...	...	...	...	...	...	.0210	.0266	.0329	.0399	.0476	

For a .875 diameter shaft with a 3/16 keyseat (E), dimension M is .0102. Using the formula given:
$J = S - (M + D) + C$; insert known values to determine J.

11

221

SECTION 11

Unit A: Plain Bearings pages 2314–2363

Plain Bearings prevent wear by providing sliding contact between mating surfaces. This unit describes the three classes of Plain Bearings and gives characteristics, applications, advantages, and disadvantages of *Plain Bearings*.

Index navigation paths and key words:

If you can't find your subject in the Primary Index, think of another key word to describe the subject.

Unit B: Ball, Roller, and Needle Bearings pages 2365–2418

Ball, roller, and needle bearings use a rolling element to carry a load while reducing friction. This unit identifies types of anti-friction bearings and their applications.

Index navigation paths and key words:

Unit C: Lubrication pages 2420–2441

The Lubrication unit begins with theories and definitions regarding friction, the physics of bodies at rest and in motion, and lubrication film. Lubricants described in this unit are of various types and compositions including synthetic based oils. The selection, application, delivery method, and contamination control are also explained.

Index navigation paths and key words:

11

Machine Elements

For more information on *interference fits* for couplings. see Section 4, Dimensioning, Gaging, and Measuring; Unit B under ANSI Standard Limits and Fits page 633.

11

223

SECTION 11

Unit H: Ball and Acme Leadscrews pages 2561–2564

Leadscrews transmit rotary motion to linear travel by using a ball screw or acme thread. This unit describes the advantages and disadvantages of each system, and applications for leadscrew assemblies.

 Index navigation paths and key words:

Unit I: Electric Motors pages 2565–2578

Machine tools are powered by electric motors. Important aspects of electric motors such as mounting dimensions, frame size, and the position of the motor shaft are controlled by the National Electrical Manufacturers Association (NEMA). This unit has a comprehensive table showing characteristics and applications of both DC and AC electric motors. A section on electric motor maintenance is also included.

 Index navigation paths and key words:

 Information on electric motor keys and keyseats is found in Section 11, Unit E: Keys and Keyseats pages 2460–2482

Unit J: Adhesives and Sealants pages 2580–2586

Joining material with adhesives offers benefits that cannot be achieved with mechanical methods. Bonded joints distribute the load over an area rather than a point, and serves as a seal. Included in this unit are mix and non-mix adhesives, threadlocking adhesives, and tapered pipe-thread sealants.

11

Safety First

Chemicals used in the shop are required to have a Material Safety Data Sheet (MSDS)

A Material Safety Data Sheet (MSDS) is designed to provide both workers and emergency personnel with the proper procedures for handling or working with a particular substance. MSDSs include information such as physical data (*melting point*, *boiling point*, *flash point* etc.), *toxicity*, health effects, first aid, reactivity, storage, disposal, *protective equipment*, and spill/leak procedures.

Index navigation paths and key words:

Information on bonding plastics is found in Section 3, Properties, Treatment, and Testing of Materials, MH29 page 593

Unit K: O-Rings pages 2588–2592

An O-ring is a one-piece flexible seal with a circular cross-section. When properly installed in a groove, an O-ring is slightly deformed so that the round cross-section is squeezed to provide a seal between two surfaces. The width and depth of the O-ring groove is critical. Check this unit for groove sizes, O-ring compounds, and installation data.

Index navigation paths and key words:

11

SECTION 11

Unit L: Rolled Steel, Wire, and Sheet Metal pages 2593–2621

Steel sections of various shapes are made by "rolling" the material between dies to produce I-beams, channels, and angles. Descriptions and sizes of rolled steel sections are standardized through a joint effort between the American Iron and Steel Institute (AISI) and the American Institute of Steel Construction (AISC). See Tables 1–7 on pages 2593–2603 for sizes.

Index navigation paths and key words:

— steel/sheet, standard gage/pages 2608–2609
— steel/structural shapes/steel/pages 2593–2602
— steel/rolled sections/shape designations/page 2593

Unit M: Shaft Alignment pages 2622–2649

What is the best way to align the shafts of two machines? Unit M illustrates a variety of procedures using dial indicators to correct the two conditions of shaft misalignment, *angularity* and *offset*.

Index navigation paths and key words:

— shaft alignment/pages 2622–2650
— shaft alignment/procedure/page 2628

Information on shaft conditions is found in Section 4, Measuring Instruments and Inspection Methods; Checking Shaft Condition MH29 pages 688–692

ASSIGNMENT

List the key terms and give a definition of each:

Keyseat	Broach	Keyseat alignment tolerance
Keyway	Keyseating machine	Clearance fit
Woodruff key	Wire EDM	Interference fit

11

APPLY IT!

Choose the correct answer.

1. What size square key is recommended for the following shaft diameters?
   a. 1/2
   b. 7/8
   c. 1 1/4
   d. 2 1/2

2. What is the width and theoretical diameters of the following Woodruff key numbers?
   a. 202
   b. 605
   c. 1208
   d. 1210

3. How high does the Woodruff key protrude above the shaft on the following assemblies?
   a. #807 key in a 1″ diameter shaft
   b. 5/16 × 1.0 1 1/4″ diameter shaft
   c. 7/8″ diameter shaft with a 3/16″ wide key

4. What size set screws should be used over the following keyed assemblies?
   a. 1 3/4″ diameter shaft with a 1/2″ wide key
   b. 2 3/8″ diameter shaft with the recommended size square key
   c. 1.625″ diameter shaft with the recommended key

5. When are rectangular keys recommended?

11

Section **12**

MEASURING UNITS

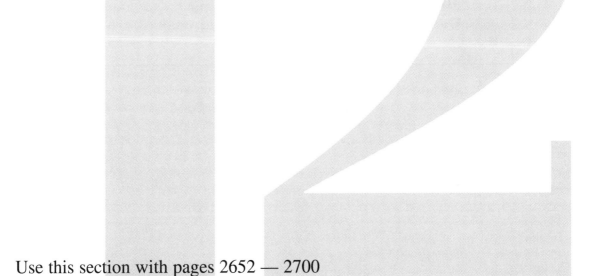

Use this section with pages 2652 — 2700
in Machinery's Handbook 29th Edition

Navigation Overview

Units Covered in this Section
>Unit B   Measuring Units
>Unit C   U.S. System and Metric System
>   Conversions

Units Covered in this Section with:
Navigation Assistant

The Navigation Assistant helps find information in
the MH29 Primary Index. The Primary Index is located
in the back of the book on pages 2701–2788 and is set
up alphabetically by subject. Watch for the magnifying
glass throughout the section for navigation hints.

May I help you?
>Unit A   Symbols and Abbreviations

Key Terms	
Meter	Minute
Millimeter	Second
Degree	Roughness Average

Measuring Units

Learning Objectives

After studying this unit you should be able to:

- Convert millimeters to inches and inches to millimeters.
- Convert degrees, minutes, and seconds to decimal degrees.
- Convert a metric–based surface finish to inches.

Introduction

The Measuring Units section of *Machinery's Handbook 29* has definitions of symbols and abbreviations, mathematical signs, and conversion tables. The unit of measurement in the United States is the *inch (in.)*. The system used for almost all of the rest of the world is the metric system. The unit of measurement for the metric system is the *meter*. When shops that use the inch or "English" system encounter engineering drawings that use the metric system, the values are converted to inches. Charts or calculators are used to convert the dimensions. It is a common practice to record the conversion directly on the drawing next to the metric dimension. Any such notes should be initialed and dated.

In the metric system, units of different sizes are changed by multiplying or dividing a single base value by powers of ten. These changes can be made simply by adding zeros or shifting the decimal point. For example, the *meter* is the basic unit of length; the kilometer is a multiple of 1000 (1000 meters); and the millimeter is a sub-multiple (one thousandth of a meter). As a result, the distances between cities and the diameter of a human hair are both measured using the meter as a standard of measurement, only the decimal point is shifted.

Despite the convenience and practicality of the metric system, the inch system with its clumsy fraction-to-decimal translations are deeply rooted in manufacturing in the United States and are not likely to change any time soon.

 Index navigation path and key words:

— metric/systems of measurement/page 2656

Unit B: Measuring Units Pages 2656–2660

The meter is equal to 39.37 inches. Metric engineering drawings use the *millimeter* (mm) as a unit of length, so one *millimeter* (0.001 meter) is equal to 0.03937 inches. The reciprocal of 0.03937 is 25.4. Use this value when converting dimensions with a calculator. For example, 1 mm ÷ 25.4 = .03937 in. One inch × 25.4 = 25.4 mm.

12

SECTION 12

Unit C: Page 2661–2700 U.S. System and Metric System Conversions

Shop Recommended page 2661

U.S. SYSTEM AND METRIC SYSTEM CONVERSIONS

Units of Length

Table 1. Linear Measure Conversion Factors

Metric	US Customary

Metric

1 kilometer (km) =
1000 meters
100,000 centimeters
1,000,000 millimeters
0.539956 nautical mile
0.621371 mile
1093.61 yards
3280.83 feet
39,370.08 inches

1 meter (m) =
10 decimeters
100 centimeters
1000 millimeters
1.09361 yards
3.28084 feet
39.37008 inches

1 decimeter (dm) = **10** centimeters

1 centimeter (cm) =
0.01 meter
10 millimeters
0.0328 foot
0.3937 inch

The millimeter (mm) is the unit of length used in metric drawings

1 millimeter (mm) =
0.001 meter
0.1 centimeter
1000 micron
0.03937 inch

1 micrometer or micron (μm) =
0.000001 meter = one millionth meter
0.0001 centimeter
0.001 millimeter
0.00003937 inch
39.37 micro-inches

US Customary

1 mile (mi) =
0.868976 nautical mile
1760 yards
5280 feet
63,360 inches
1.609344 kilometers
1609.344 meters
160,934.4 centimeters
1,609,344 millimeters

1 yard (yd) =
3 feet
36 inches
0.9144 meter
91.44 centimeter
914.4 millimeter

1 foot (international) (ft) =
12 inches = ⅓ yard
0.3048 meter
30.48 centimeter
304.8 millimeters

1 survey foot =
1.000002 international feet
$^{12}/_{39.37}$ = 0.3048006096012 meter

1 inch (in) =
1000 mils
1,000,000 micro-inch
2.54 centimeters
25.4 millimeters
25,400 microns

One inch is equal to 25.4 mm.

1 mil =
0.001 inch
1000 micro-inches
0.0254 millimeters

1 micro-inch (μin) =
0.000001 inch = one millionth inch
0.0254 micrometer (micron)

12

232

Shop Recommended page 2662, Table 2

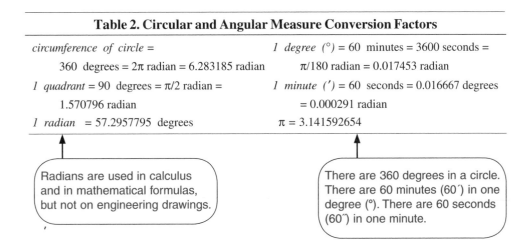

Table 2. Circular and Angular Measure Conversion Factors

circumference of circle =
360 degrees = 2π radian = 6.283185 radian

1 quadrant = 90 degrees = $\pi/2$ radian =
1.570796 radian

1 radian = 57.2957795 degrees

1 degree (°) = 60 minutes = 3600 seconds =
$\pi/180$ radian = 0.017453 radian

1 minute (′) = 60 seconds = 0.016667 degrees
= 0.000291 radian

$\pi = 3.141592654$

> Radians are used in calculus and in mathematical formulas, but not on engineering drawings.

> There are 360 degrees in a circle. There are 60 minutes (60′) in one degree (°). There are 60 seconds (60″) in one minute.

Degrees are very often a value that needs to be broken down into smaller units. *Degrees, minutes*, and *seconds* use a system that uses a base of 60, not unlike a clock. One degree (hour) is equal to 60 minutes (60′). There are 60 *seconds* (60″) in one *minute*. *Degrees, minutes* and *seconds* are difficult to add, subtract, multiply and divide. For this reason they are converted to decimal degrees.

Angular Conversions

To convert an angle from *degrees, minutes*, and *seconds* to decimal degrees, divide by 60:

$$30' \div 60 = .5°$$
$$10° \ 15' = 10 + (15 \div 60) = 10.25°$$

For conversions involving minutes and seconds begin with the seconds. For example, for the angle 14° 35′27″ begin with 27″:

$$27 \text{ (seconds)} \div 60 = .45$$
$$.45 + 35 \text{ (minutes)} = 35.45$$
$$35.45 \div 60 = .59083333°$$
$$\text{Add } 14°$$
$$14° \ 35'27'' = 14.59083333°$$

Scientific calculators can easily convert *degrees, minutes, and seconds* to *decimal degrees*. Look for the DMS ➝ DD key.

12

Shop Recommended page 2664, Table 7

Table 7. Fractional Inch to Decimal Inch and Millimeter

Fractional Inch	Decimal Inch	Millimeters	Fractional Inch	Decimal Inch	Millimeters
1/64	0.015625	0.396875		0.511811024	13
1/32	0.03125	0.79375	33/64	0.515625	13.096875
	0.039370079	1	17/32	0.53125	13.49375
3/64	0.046875	1.190625	35/64	0.546875	13.890625
1/16	0.0625	1.5875		0.551181102	14
5/64	0.078125	1.984375	9/16	0.5625	14.2875
	0.078740157	2	37/64	0.578125	14.684375
3/32	0.09375	2.38125		0.590551181	15
7/64	0.109375	2.778125	19/32	0.59375	15.08125
	0.118110236	3	39/64	0.609375	15.478125
1/8	0.125	3.175	5/8	0.625	15.875
9/64	0.140625	3.571875		0.62992126	16
5/32	0.15625	3.96875	41/64	0.640625	16.271875
	0.157480315	4	21/32	0.65625	16.66875
11/64	0.171875	4.365625		0.669291339	17
3/16	0.1875	4.7625	43/64	0.671875	17.065625
	0.196850394	5	11/16	0.6875	17.4625
13/64	0.203125	5.159375	45/64	0.703125	17.859375
7/32	0.21875	5.55625		0.708661417	18
15/64	0.234375	5.953125	23/32	0.71875	18.25625
	0.236220472	6	47/64	0.734375	18.653125
1/4	0.25	6.35		0.748031496	19
17/64	0.265625	6.746875	3/4	0.75	19.05
	0.275590551	7	49/64	0.765625	19.446875
9/32	0.28125	7.14375	25/32	0.78125	19.84375
19/64	0.296875	7.540625		0.787401575	20
5/16	0.3125	7.9375	51/64	0.796875	20.240625
	0.31496063	8	13/16	0.8125	20.6375
21/64	0.328125	8.334375		0.826771654	21
11/32	0.34375	8.73125	53/64	0.828125	21.034375
	0.354330709	9	27/32	0.84375	21.43125
23/64	0.359375	9.128125	55/64	0.859375	21.828125
3/8	0.375	9.525		0.866141732	22
25/64	0.390625	9.921875	7/8	0.875	22.225
	0.393700787	10	57/64	0.890625	22.621875
13/32	0.40625	10.31875		0.905511811	23
27/64	0.421875	10.715625	29/32	0.90625	23.01875
	0.433070866	11	59/64	0.921875	23.415625
7/16	0.4375	11.1125	15/16	0.9375	23.8125
29/64	0.453125	11.509375		0.94488189	24
15/32	0.46875	11.90625	61/64	0.953125	24.209375
	0.472440945	12	31/32	0.96875	24.60625
31/64	0.484375	12.303125		0.984251969	25
1/2	0.5	12.7	63/64	0.984375	25.003125

Shop Recommended page 2665, Tables 8a and 8b

Table 8a. Inch to Millimeters Conversion

inch	mm	inch	mm	inch	mm	inch	mm	inch	mm	inch	mm
10	254.00000	1	25.40000	0.1	2.54000	.01	0.25400	0.001	0.02540	0.0001	0.00254
20	508.00000	2	50.80000	0.2	5.08000	.02	0.50800	0.002	0.05080	0.0002	0.00508
30	762.00000	3	76.20000	0.3	7.62000	.03	0.76200	0.003	0.07620	0.0003	0.00762
40	1,016.00000	4	101.60000	0.4	10.16000	.04	1.01600	0.004	0.10160	0.0004	0.01016
50	1,270.00000	5	127.00000	0.5	12.70000	.05	1.27000	0.005	0.12700	0.0005	0.01270
60	1,524.00000	6	152.40000	0.6	15.24000	.06	1.52400	0.006	0.15240	0.0006	0.01524
70	1,778.00000	7	177.80000	0.7	17.78000	.07	1.77800	0.007	0.17780	0.0007	0.01778
80	2,032.00000	8	203.20000	0.8	20.32000	.08	2.03200	0.008	0.20320	0.0008	0.02032
90	2,286.00000	9	228.60000	0.9	22.86000	.09	2.2860	0.009	0.22860	0.0009	0.02286
100	2,540.00000	10	254.00000	1.0	25.40000	.10	2.54000	0.010	0.25400	0.0010	0.02540

All values in this table are exact. For inches to centimeters, shift decimal point in mm column one place to left and read centimeters, thus, for example, 40 in. = 1016 mm = 101.6 cm.

Table 8b. Millimeters to Inch Conversion

mm	inch	mm	inch	mm	inch	mm	inch	mm	inch	mm	inch
100	3.93701	10	0.39370	1	0.03937	0.1	0.00394	0.01	.000039	0.001	0.00004
200	7.87402	20	0.78740	2	0.07874	0.2	0.00787	0.02	.00079	0.002	0.00008
300	11.81102	30	1.18110	3	0.11811	0.3	0.01181	0.03	.00118	0.003	0.00012
400	15.74803	40	1.57480	4	0.15748	0.4	0.01575	0.04	.00157	0.004	0.00016
500	19.68504	50	1.96850	5	0.19685	0.5	0.01969	0.05	.00197	0.005	0.00020
600	23.62205	60	2.36220	6	0.23622	0.6	0.02362	0.06	.00236	0.006	0.00024
700	27.55906	70	2.75591	7	0.27559	0.7	0.02756	0.07	.00276	0.007	0.00028
800	31.49606	80	3.14961	8	0.31496	0.8	0.03150	0.08	.00315	0.008	0.00031
900	35.43307	90	3.54331	9	0.35433	0.9	0.03543	0.09	.00354	0.009	0.00035
1,000	39.37008	100	3.93701	10	0.39370	1.0	0.03937	0.10	.00394	0.010	0.00039

Based on 1 inch = 25.4 millimeters, exactly. For centimeters to inches, shift decimal point of centimeter value one place to right and enter mm column, thus, for example, 70 cm = 700 mm = 27.55906 inches.

A mixed fractional inch combines a whole number with a common fraction. To use the following type of table, find the fraction in the left column and the whole number in the top row.

Example
Convert 4 5/16 to millimeters.

12

235

Shop Recommended page 2666, Table 10

4 and 5/16 = 109.5375 mm

Table 10. Mixed Fractional Inches to Millimeters Conversion for 0 to 41 Inches in $\frac{1}{64}$-Inch Increments

Inches↓	0	1	2	3	4	5	6	7	8	9	10	20	30	40
							Millimeters							
0	0	25.4	50.8	76.2	101.6	127.0	152.4	177.8	203.2	228.6	254.0	508.0	762.0	1016.0
1/64	0.396875	25.796875	51.196875	76.596875	101.996875	127.396875	152.796875	178.196875	203.596875	228.996875	254.396875	508.396875	762.396875	1016.396875
1/32	0.79375	26.19375	51.59375	76.99375	102.39375	127.79375	153.19375	178.59375	203.99375	229.39375	254.79375	508.79375	762.79375	1016.79375
3/64	1.190625	26.590625	51.990625	77.390625	102.790625	128.190625	153.590625	178.990625	204.390625	229.790625	255.190625	509.190625	763.190625	1017.190625
1/16	1.5875	26.9875	52.3875	77.7875	103.1875	128.5875	153.9875	179.3875	204.7875	230.1875	255.5875	509.5875	763.5875	1017.5875
5/64	1.984375	27.384375	52.784375	78.184375	103.584375	128.984375	154.384375	179.784375	205.184375	230.584375	255.984375	509.984375	763.984375	1017.984375
3/32	2.38125	27.78125	53.18125	78.58125	103.98125	129.38125	154.78125	180.18125	205.58125	230.98125	256.38125	510.38125	764.38125	1018.38125
7/64	2.778125	28.178125	53.578125	78.978125	104.378125	129.778125	155.178125	180.578125	205.978125	231.378125	256.778125	510.778125	764.778125	1018.778125
1/8	3.175	28.575	53.975	79.375	104.775	130.175	155.575	180.975	206.375	231.775	257.175	511.175	765.175	1019.175
9/64	3.571875	28.971875	54.371875	79.771875	105.171875	130.571875	155.971875	181.371875	206.771875	232.171875	257.571875	511.571875	765.571875	1019.571875
5/32	3.96875	29.36875	54.76875	80.16875	105.56875	130.96875	156.36875	181.76875	207.16875	232.56875	257.96875	511.96875	765.96875	1019.96875
11/64	4.365625	29.765625	55.165625	80.565625	105.965625	131.365625	156.765625	182.165625	207.565625	232.965625	258.365625	512.365625	766.365625	1020.365625
3/16	4.7625	30.1625	55.5625	80.9625	106.3625	131.7625	157.1625	182.5625	207.9625	233.3625	258.7625	512.7625	766.7625	1020.7625
13/64	5.159375	30.559375	55.959375	81.359375	106.759375	132.159375	157.559375	182.959375	208.359375	233.759375	259.159375	513.159375	767.159375	1021.159375
7/32	5.55625	30.95625	56.35625	81.75625	107.15625	132.55625	157.95625	183.35625	208.75625	234.15625	259.55625	513.55625	767.55625	1021.55625
15/64	5.953125	31.353125	56.753125	82.153125	107.553125	132.953125	158.353125	183.753125	209.153125	234.553125	259.953125	513.953125	767.953125	1021.953125
1/4	6.35	31.75	57.15	82.55	107.95	133.35	158.75	184.15	209.55	234.95	260.35	514.35	768.35	1022.35
17/64	6.746875	32.146875	57.546875	82.946875	108.346875	133.746875	159.146875	184.546875	209.946875	235.346875	260.746875	514.746875	768.746875	1022.746875
9/32	7.14375	32.54375	57.94375	83.34375	108.74375	134.14375	159.54375	184.94375	210.34375	235.74375	261.14375	515.14375	769.14375	1023.14375
19/64	7.540625	32.940625	58.340625	83.740625	109.140625	134.540625	159.940625	185.340625	210.740625	236.140625	261.540625	515.540625	769.540625	1023.540625
5/16	7.9375	33.3375	58.7375	84.1375	109.5375	134.9375	160.3375	185.7375	211.1375	236.5375	261.9375	515.9375	769.9375	1023.9375
21/64	8.334375	33.734375	59.134375	84.534375	109.934375	135.334375	160.734375	186.134375	211.534375	236.934375	262.334375	516.334375	770.334375	1024.334375
11/32	8.73125	34.13125	59.53125	84.93125	110.33125	135.73125	161.13125	186.53125	211.93125	237.33125	262.73125	516.73125	770.73125	1024.73125
23/64	9.128125	34.528125	59.928125	85.328125	110.728125	136.128125	161.528125	186.928125	212.328125	237.728125	263.128125	517.128125	771.128125	1025.128125
3/8	9.525	34.925	60.325	85.725	111.125	136.525	161.925	187.325	212.725	238.125	263.525	517.525	771.525	1025.525
25/64	9.921875	35.321875	60.721875	86.121875	111.521875	136.921875	162.321875	187.721875	213.121875	238.521875	263.921875	517.921875	771.921875	1025.921875
13/32	10.31875	35.71875	61.11875	86.51875	111.91875	137.31875	162.71875	188.11875	213.51875	238.91875	264.31875	518.31875	772.31875	1026.31875
27/64	10.715625	36.115625	61.515625	86.915625	112.315625	137.715625	163.115625	188.515625	213.915625	239.315625	264.715625	518.715625	772.715625	1026.715625
7/16	11.1125	36.5125	61.9125	87.3125	112.7125	138.1125	163.5125	188.9125	214.3125	239.7125	265.1125	519.1125	773.1125	1027.1125
29/64	11.509375	36.909375	62.309375	87.709375	113.109375	138.509375	163.909375	189.309375	214.709375	240.109375	265.509375	519.509375	773.509375	1027.509375
15/32	11.90625	37.30625	62.70625	88.10625	113.50625	138.90625	164.30625	189.70625	215.10625	240.50625	265.90625	519.90625	773.90625	1027.90625
31/64	12.303125	37.703125	63.103125	88.503125	113.903125	139.303125	164.703125	190.103125	215.503125	240.903125	266.303125	520.303125	774.303125	1028.303125
1/2	12.7	38.1	63.5	88.9	114.3	139.7	165.1	190.5	215.9	241.3	266.7	520.7	774.7	1028.7

12

Shop Recommended page 2668, Table 11

Table 11. Decimals of an Inch to Millimeters Conversion

<table>
<tr><td>→
Inches
↓</td><td>0.000</td><td>0.001</td><td>0.002</td><td>0.003</td><td>0.004</td><td>0.005</td><td>0.006</td><td>0.007</td><td>0.008</td><td>0.009</td></tr>
<tr><td></td><td colspan="10" align="center">Millimeters</td></tr>
<tr><td>0.000</td><td>…</td><td>0.0254</td><td>0.0508</td><td>0.0762</td><td>0.1016</td><td>0.1270</td><td>0.1524</td><td>0.1778</td><td>0.2032</td><td>0.2286</td></tr>
<tr><td>0.010</td><td>0.2540</td><td>0.2794</td><td>0.3048</td><td>0.3302</td><td>0.3556</td><td>0.3810</td><td>0.4064</td><td>0.4318</td><td>0.4572</td><td>0.4826</td></tr>
<tr><td>0.020</td><td>0.5080</td><td>0.5334</td><td>0.5588</td><td>0.5842</td><td>0.6096</td><td>0.6350</td><td>0.6604</td><td>0.6858</td><td>0.7112</td><td>0.7366</td></tr>
<tr><td>0.030</td><td>0.7620</td><td>0.7874</td><td>0.8128</td><td>0.8382</td><td>0.8636</td><td>0.8890</td><td>0.9144</td><td>0.9398</td><td>0.9652</td><td>0.9906</td></tr>
<tr><td>0.040</td><td>1.0160</td><td>1.0414</td><td>1.0668</td><td>1.0922</td><td>1.1176</td><td>1.1430</td><td>1.1684</td><td>1.1938</td><td>1.2192</td><td>1.2446</td></tr>
<tr><td>0.050</td><td>1.2700</td><td>1.2954</td><td>1.3208</td><td>1.3462</td><td>1.3716</td><td>1.3970</td><td>1.4224</td><td>1.4478</td><td>1.4732</td><td>1.4986</td></tr>
<tr><td>0.060</td><td>1.5240</td><td>1.5494</td><td>1.5748</td><td>1.6002</td><td>1.6256</td><td>1.6510</td><td>1.6764</td><td>1.7018</td><td>1.7272</td><td>1.7526</td></tr>
<tr><td>0.070</td><td>1.7780</td><td>1.8034</td><td>1.8288</td><td>1.8542</td><td>1.8796</td><td>1.9050</td><td>1.9304</td><td>1.9558</td><td>1.9812</td><td>2.0066</td></tr>
<tr><td>0.080</td><td>2.0320</td><td>2.0574</td><td>2.0828</td><td>2.1082</td><td>2.1336</td><td>2.1590</td><td>2.1844</td><td>2.2098</td><td>2.2352</td><td>2.2606</td></tr>
<tr><td>0.090</td><td>2.2860</td><td>2.3114</td><td>2.3368</td><td>2.3622</td><td>2.3876</td><td>2.4130</td><td>2.4384</td><td>2.4638</td><td>2.4892</td><td>2.5146</td></tr>
<tr><td>0.100</td><td>2.5400</td><td>2.5654</td><td>2.5908</td><td>2.6162</td><td>2.6416</td><td>2.6670</td><td>2.6924</td><td>2.7178</td><td>2.7432</td><td>2.7686</td></tr>
<tr><td>0.110</td><td>2.7940</td><td>2.8194</td><td>2.8448</td><td>2.8702</td><td>2.8956</td><td>2.9210</td><td>2.9464</td><td>2.9718</td><td>2.9972</td><td>3.0226</td></tr>
<tr><td>0.120</td><td>3.0480</td><td>3.0734</td><td>3.0988</td><td>3.1242</td><td>3.1496</td><td>3.1750</td><td>3.2004</td><td>3.2258</td><td>3.2512</td><td>3.2766</td></tr>
<tr><td>0.130</td><td>3.3020</td><td>3.3274</td><td>3.3528</td><td>3.3782</td><td>3.4036</td><td>3.4290</td><td>3.4544</td><td>3.4798</td><td>3.5052</td><td>3.5306</td></tr>
<tr><td>0.140</td><td>3.5560</td><td>3.5814</td><td>3.6068</td><td>3.6322</td><td>3.6576</td><td>3.6830</td><td>3.7084</td><td>3.7338</td><td>3.7592</td><td>3.7846</td></tr>
<tr><td>0.150</td><td>3.8100</td><td>3.8354</td><td>3.8608</td><td>3.8862</td><td>3.9116</td><td>3.9370</td><td>3.9624</td><td>3.9878</td><td>4.0132</td><td>4.0386</td></tr>
<tr><td>0.160</td><td>4.0640</td><td>4.0894</td><td>4.1148</td><td>4.1402</td><td>4.1656</td><td>4.1910</td><td>4.2164</td><td>4.2418</td><td>4.2672</td><td>4.2926</td></tr>
<tr><td>0.170</td><td>4.3180</td><td>4.3434</td><td>4.3688</td><td>4.3942</td><td>4.4196</td><td>4.4450</td><td>4.4704</td><td>4.4958</td><td>4.5212</td><td>4.5466</td></tr>
<tr><td>0.180</td><td>4.5720</td><td>4.5974</td><td>4.6228</td><td>4.6482</td><td>4.6736</td><td>4.6990</td><td>4.7244</td><td>4.7498</td><td>4.7752</td><td>4.8006</td></tr>
<tr><td>0.190</td><td>4.8260</td><td>4.8514</td><td>4.8768</td><td>4.9022</td><td>4.9276</td><td>4.9530</td><td>4.9784</td><td>5.0038</td><td>5.0292</td><td>5.0546</td></tr>
<tr><td>0.200</td><td>5.0800</td><td>5.1054</td><td>5.1308</td><td>5.1562</td><td>5.1816</td><td>5.2070</td><td>5.2324</td><td>5.2578</td><td>5.2832</td><td>5.3086</td></tr>
<tr><td>0.210</td><td>5.3340</td><td>5.3594</td><td>5.3848</td><td>5.4102</td><td>5.4356</td><td>5.4610</td><td>5.4864</td><td>5.5118</td><td>5.5372</td><td>5.5626</td></tr>
<tr><td>0.220</td><td>5.5880</td><td>5.6134</td><td>5.6388</td><td>5.6642</td><td>5.6896</td><td>5.7150</td><td>5.7404</td><td>5.7658</td><td>5.7912</td><td>5.8166</td></tr>
<tr><td>0.230</td><td>5.8420</td><td>5.8674</td><td>5.8928</td><td>5.9182</td><td>5.9436</td><td>5.9690</td><td>5.9944</td><td>6.0198</td><td>6.0452</td><td>6.0706</td></tr>
<tr><td>0.240</td><td>6.0960</td><td>6.1214</td><td>6.1468</td><td>6.1722</td><td>6.1976</td><td>6.2230</td><td>6.2484</td><td>6.2738</td><td>6.2992</td><td>6.3246</td></tr>
<tr><td>0.250</td><td>6.3500</td><td>6.3754</td><td>6.4008</td><td>6.4262</td><td>6.4516</td><td>6.4770</td><td>6.5024</td><td>6.5278</td><td>6.5532</td><td>6.5786</td></tr>
<tr><td>0.260</td><td>6.6040</td><td>6.6294</td><td>6.6548</td><td>6.6802</td><td>6.7056</td><td>6.7310</td><td>6.7564</td><td>6.7818</td><td>6.8072</td><td>6.8326</td></tr>
<tr><td>0.270</td><td>6.8580</td><td>6.8834</td><td>6.9088</td><td>6.9342</td><td>6.9596</td><td>6.9850</td><td>7.0104</td><td>7.0358</td><td>7.0612</td><td>7.0866</td></tr>
<tr><td>0.280</td><td>7.1120</td><td>7.1374</td><td>7.1628</td><td>7.1882</td><td>7.2136</td><td>7.2390</td><td>7.2644</td><td>7.2898</td><td>7.3152</td><td>7.3406</td></tr>
<tr><td>0.290</td><td>7.3660</td><td>7.3914</td><td>7.4168</td><td>7.4422</td><td>7.4676</td><td>7.4930</td><td>7.5184</td><td>7.5438</td><td>7.5692</td><td>7.5946</td></tr>
<tr><td>0.300</td><td>7.6200</td><td>7.6454</td><td>7.6708</td><td>7.6962</td><td>7.7216</td><td>7.7470</td><td>7.7724</td><td>7.7978</td><td>7.8232</td><td>7.8486</td></tr>
<tr><td>0.310</td><td>7.8740</td><td>7.8994</td><td>7.9248</td><td>7.9502</td><td>7.9756</td><td>8.0010</td><td>8.0264</td><td>8.0518</td><td>8.0772</td><td>8.1026</td></tr>
<tr><td>0.320</td><td>8.1280</td><td>8.1534</td><td>8.1788</td><td>8.2042</td><td>8.2296</td><td>8.2550</td><td>8.2804</td><td>8.3058</td><td>8.3312</td><td>8.3566</td></tr>
<tr><td>0.330</td><td>8.3820</td><td>8.4074</td><td>8.4328</td><td>8.4582</td><td>8.4836</td><td>8.5090</td><td>8.5344</td><td>8.5598</td><td>8.5852</td><td>8.6106</td></tr>
<tr><td>0.340</td><td>8.6360</td><td>8.6614</td><td>8.6868</td><td>8.7122</td><td>8.7376</td><td>8.7630</td><td>8.7884</td><td>8.8138</td><td>8.8392</td><td>8.8646</td></tr>
<tr><td>0.350</td><td>8.8900</td><td>8.9154</td><td>8.9408</td><td>8.9662</td><td>8.9916</td><td>9.0170</td><td>9.0424</td><td>9.0678</td><td>9.0932</td><td>9.1186</td></tr>
<tr><td>0.360</td><td>9.1440</td><td>9.1694</td><td>9.1948</td><td>9.2202</td><td>9.2456</td><td>9.2710</td><td>9.2964</td><td>9.3218</td><td>9.3472</td><td>9.3726</td></tr>
<tr><td>0.370</td><td>9.3980</td><td>9.4234</td><td>9.4488</td><td>9.4742</td><td>9.4996</td><td>9.5250</td><td>9.5504</td><td>9.5758</td><td>9.6012</td><td>9.6266</td></tr>
<tr><td>0.380</td><td>9.6520</td><td>9.6774</td><td>9.7028</td><td>9.7282</td><td>9.7536</td><td>9.7790</td><td>9.8044</td><td>9.8298</td><td>9.8552</td><td>9.8806</td></tr>
<tr><td>0.390</td><td>9.9060</td><td>9.9314</td><td>9.9568</td><td>9.9822</td><td>10.0076</td><td>10.0330</td><td>10.0584</td><td>10.0838</td><td>10.1092</td><td>10.1346</td></tr>
<tr><td>0.400</td><td>10.1600</td><td>10.1854</td><td>10.2108</td><td>10.2362</td><td>10.2616</td><td>10.2870</td><td>10.3124</td><td>10.3378</td><td>10.3632</td><td>10.3886</td></tr>
<tr><td>0.410</td><td>10.4140</td><td>10.4394</td><td>10.4648</td><td>10.4902</td><td>10.5156</td><td>10.5410</td><td>10.5664</td><td>10.5918</td><td>10.6172</td><td>10.6426</td></tr>
<tr><td>0.420</td><td>10.6680</td><td>10.6934</td><td>10.7188</td><td>10.7442</td><td>10.7696</td><td>10.7950</td><td>10.8204</td><td>10.8458</td><td>10.8712</td><td>10.8966</td></tr>
<tr><td>0.430</td><td>10.9220</td><td>10.9474</td><td>10.9728</td><td>10.9982</td><td>11.0236</td><td>11.0490</td><td>11.0744</td><td>11.0998</td><td>11.1252</td><td>11.1506</td></tr>
<tr><td>0.440</td><td>11.1760</td><td>11.2014</td><td>11.2268</td><td>11.2522</td><td>11.2776</td><td>11.3030</td><td>11.3284</td><td>11.3538</td><td>11.3792</td><td>11.4046</td></tr>
<tr><td>0.450</td><td>11.4300</td><td>11.4554</td><td>11.4808</td><td>11.5062</td><td>11.5316</td><td>11.5570</td><td>11.5824</td><td>11.6078</td><td>11.6332</td><td>11.6586</td></tr>
<tr><td>0.460</td><td>11.6840</td><td>11.7094</td><td>11.7348</td><td>11.7602</td><td>11.7856</td><td>11.8110</td><td>11.8364</td><td>11.8618</td><td>11.8872</td><td>11.9126</td></tr>
<tr><td>0.470</td><td>11.9380</td><td>11.9634</td><td>11.9888</td><td>12.0142</td><td>12.0396</td><td>12.0650</td><td>12.0904</td><td>12.1158</td><td>12.1412</td><td>12.1666</td></tr>
<tr><td>0.480</td><td>12.1920</td><td>12.2174</td><td>12.2428</td><td>12.2682</td><td>12.2936</td><td>12.3190</td><td>12.3444</td><td>12.3698</td><td>12.3952</td><td>12.4206</td></tr>
<tr><td>0.490</td><td>12.4460</td><td>12.4714</td><td>12.4968</td><td>12.5222</td><td>12.5476</td><td>12.5730</td><td>12.5984</td><td>12.6238</td><td>12.6492</td><td>12.6746</td></tr>
<tr><td>0.500</td><td>12.7000</td><td>12.7254</td><td>12.7508</td><td>12.7762</td><td>12.8016</td><td>12.8270</td><td>12.8524</td><td>12.8778</td><td>12.9032</td><td>12.9286</td></tr>
</table>

To convert .125 to mm, find .120 in the left column and .005 in the top row.

Follow the line until it crosses the .005 column. .125 = 3.175mm

Table 12 on page 2670 converts millimeters to inches. To use the table to convert 83mm to inches, find "80" in the left column and "3" in the top row. The intersecting value is 3.26772 inches.

12

237

Shop Recommended page 2670, Table 12

Table 12. Millimeters to Inches Conversion

→ Millimeters ↓	0	1	2	3	4	5	6	7	8	9
					Inches					
0	...	0.03937	0.07874	0.11811	0.15748	0.19685	0.23622	0.27559	0.31496	0.35433
10	0.39370	0.43307	0.47244	0.51181	0.55118	0.59055	0.62992	0.66929	0.70866	0.74803
20	0.78740	0.82677	0.86614	0.90551	0.94488	0.98425	1.02362	1.06299	1.10236	1.14173
30	1.18110	1.22047	1.25984	1.29921	1.33858	1.37795	1.41732	1.45669	1.49606	1.53543
40	1.57480	1.61417	1.65354	1.69291	1.73228	1.77165	1.81102	1.85039	1.88976	1.92913
50	1.96850	2.00787	2.04724	2.08661	2.12598	2.16535	2.20472	2.24409	2.28346	2.32283
60	2.36220	2.40157	2.44094	2.48031	2.51969	2.55906	2.59843	2.63780	2.67717	2.71654
70	2.75591	2.79528	2.83465	2.87402	2.91339	2.95276	2.99213	3.03150	3.07087	3.11024
80	3.14961	3.18898	3.22835	3.26772	3.30709	3.34646	3.38583	3.42520	3.46457	3.50394
90	3.54331	3.58268	3.62205	3.66142	3.70079	3.74016	3.77953	3.81890	3.85827	3.89764
100	3.93701	3.97638	4.01575	4.05512	4.09449	4.13386	4.17323	4.21260	4.25197	4.29134
110	4.33071	4.37008	4.40945	4.44882	4.48819	4.52756	4.56693	4.60630	4.64567	4.68504
120	4.72441	4.76378	4.80315	4.84252	4.88189	4.92126	4.96063	5.00000	5.03937	5.07874
130	5.11811	5.15748	5.19685	5.23622	5.27559	5.31496	5.35433	5.39370	5.43307	5.47244
140	5.51181	5.55118	5.59055	5.62992	5.66929	5.70866	5.74803	5.78740	5.82677	5.86614
150	5.90551	5.94488	5.98425	6.02362	6.06299	6.10236	6.14173	6.18110	6.22047	6.25984
160	6.29921	6.33858	6.37795	6.41732	6.45669	6.49606	6.53543	6.57480	6.61417	6.65354
170	6.69291	6.73228	6.77165	6.81102	6.85039	6.88976	6.92913	6.96850	7.00787	7.04724
180	7.08661	7.12598	7.16535	7.20472	7.24409	7.28346	7.32283	7.36220	7.40157	7.44094
190	7.48031	7.51969	7.55906	7.59843	7.63780	7.67717	7.71654	7.75591	7.79528	7.83465
200	7.87402	7.91339	7.95276	7.99213	8.03150	8.07087	8.11024	8.14961	8.18898	8.22835
210	8.26772	8.30709	8.34646	8.38583	8.42520	8.46457	8.50394	8.54331	8.58268	8.62205
220	8.66142	8.70079	8.74016	8.77953	8.81890	8.85827	8.89764	8.93701	8.97638	9.01575
230	9.05512	9.09449	9.13386	9.17323	9.21260	9.25197	9.29134	9.33071	9.37008	9.40945
240	9.44882	9.48819	9.52756	9.56693	9.60630	9.64567	9.68504	9.72441	9.76378	9.80315
250	9.84252	9.88189	9.92126	9.96063	10.0000	10.0394	10.0787	10.1181	10.1575	10.1969
260	10.2362	10.2756	10.3150	10.3543	10.3937	10.4331	10.4724	10.5118	10.5512	10.5906
270	10.6299	10.6693	10.7087	10.7480	10.7874	10.8268	10.8661	10.9055	10.9449	10.9843
280	11.0236	11.0630	11.1024	11.1417	11.1811	11.2205	11.2598	11.2992	11.3386	11.3780
290	11.4173	11.4567	11.4961	11.5354	11.5748	11.6142	11.6535	11.6929	11.7323	11.7717
300	11.8110	11.8504	11.8898	11.9291	11.9685	12.0079	12.0472	12.0866	12.1260	12.1654
310	12.2047	12.2441	12.2835	12.3228	12.3622	12.4016	12.4409	12.4803	12.5197	12.5591
320	12.5984	12.6378	12.6772	12.7165	12.7559	12.7953	12.8346	12.8740	12.9134	12.9528
330	12.9921	13.0315	13.0709	13.1102	13.1496	13.1890	13.2283	13.2677	13.3071	13.3465
340	13.3858	13.4252	13.4646	13.5039	13.5433	13.5827	13.6220	13.6614	13.7008	13.7402
350	13.7795	13.8189	13.8583	13.8976	13.9370	13.9764	14.0157	14.0551	14.0945	14.1339
360	14.1732	14.2126	14.2520	14.2913	14.3307	14.3701	14.4094	14.4488	14.4882	14.5276
370	14.5669	14.6063	14.6457	14.6850	14.7244	14.7638	14.8031	14.8425	14.8819	14.9213
380	14.9606	15.0000	15.0394	15.0787	15.1181	15.1575	15.1969	15.2362	15.2756	15.3150
390	15.3543	15.3937	15.4331	15.4724	15.5118	15.5512	15.5906	15.6299	15.6693	15.7087
400	15.7480	15.7874	15.8268	15.8661	15.9055	15.9449	15.9843	16.0236	16.0630	16.1024
410	16.1417	16.1811	16.2205	16.2598	16.2992	16.3386	16.3780	16.4173	16.4567	16.4961
420	16.5354	16.5748	16.6142	16.6535	16.6929	16.7323	16.7717	16.8110	16.8504	16.8898
430	16.9291	16.9685	17.0079	17.0472	17.0866	17.1260	17.1654	17.2047	17.2441	17.2835
440	17.3228	17.3622	17.4016	17.4409	17.4803	17.5197	17.5591	17.5984	17.6378	17.6772
450	17.7165	17.7559	17.7953	17.8346	17.8740	17.9134	17.9528	17.9921	18.0315	18.0709
460	18.1102	18.1496	18.1890	18.2283	18.2677	18.3071	18.3465	18.3858	18.4252	18.4646
470	18.5039	18.5433	18.5827	18.6220	18.6614	18.7008	18.7402	18.7795	18.8189	18.8583
480	18.8976	18.9370	18.9764	19.0157	19.0551	19.0945	19.1339	19.1732	19.2126	19.2520
490	19.2913	19.3307	19.3701	19.4094	19.4488	19.4882	19.5276	19.5669	19.6063	19.6457

83mm = 3.26772 in.

12

Surface Finish

Designers specify the smoothness of a surface based on the type of service, function, and operating conditions of the part or assembly. A symbol is used to convey the "roughness" requirement of the surface. Page 743 of *Machinery's Handbook 29* shows a variety of acceptable methods for applying surface finish symbols.

Shop Recommended page 743

SURFACE TEXTURE

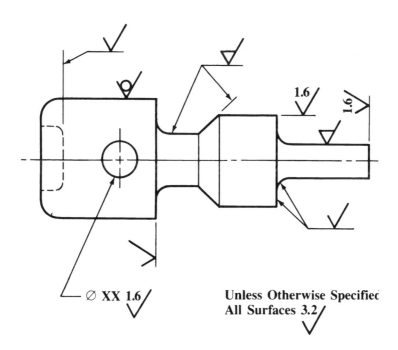

⌀ XX 1.6

**Unless Otherwise Specified
All Surfaces 3.2**

Conversion of Surface Finishes

Control of the surface texture or *finish* of a designed part is common. The roughness of a surface is given in *microinches* (μin) which is one-millionth of an inch (.000001). Metric drawings give the surface finish in *micrometers* (μm) which is one-millionth of a meter (.000001). These are measurements of the roughness of a surface known as *roughness average* or Ra. The roughness average is an average of the peaks and valleys observed in the inspection of a finish using an instrument known as a profilometer. Smaller numbers are smoother than large numbers. In the inch system, a 500 finish is as rough as a saw cut whereas a number 2 finish looks like a mirror.

The conversion of metric surface finishes to inches is sometimes a matter of confusion. Convert metric finishes to inches using Table 13b on page 2673. A number 3.2 finish found on a metric drawing is the same as a 125 finish in inches.

12

Shop Recommended page 2673, Table 13b

Use this table to convert metric surface finishes to inch surface finishes.

Table 13b. Micrometers (microns) to Microinches Conversion

→ Microns ↓	0	0.01	0.02	0.03	0.04	0.05	0.06	0.07	0.08	0.09
					Microinches					
0.00	0.0000	0.3937	0.7874	1.1811	1.5748	1.9685	2.3622	2.7559	3.1496	3.5433
0.10	3.9370	4.3307	4.7244	5.1181	5.5118	5.9055	6.2992	6.6929	7.0866	7.4803
0.20	7.8740	8.2677	8.6614	9.0551	9.4488	9.8425	10.2362	10.6299	11.0236	11.4173
0.30	11.8110	12.2047	12.5984	12.9921	13.3858	13.7795	14.1732	14.5669	14.9606	15.3543
0.40	15.7480	16.1417	16.5354	16.9291	17.3228	17.7165	18.1102	18.5039	18.8976	19.2913
0.50	19.6850	20.0787	20.4724	20.8661	21.2598	21.6535	22.0472	22.4409	22.8346	23.2283
0.60	23.6220	24.0157	24.4094	24.8031	25.1969	25.5906	25.9843	26.3780	26.7717	27.1654
0.70	27.5591	27.9528	28.3465	28.7402	29.1339	29.5276	29.9213	30.3150	30.7087	31.1024
0.80	31.4961	31.8898	32.2835	32.6772	33.0709	33.4646	33.8583	34.2520	34.6457	35.0394
0.90	35.4331	35.8268	36.2205	36.6142	37.0079	37.4016	37.7953	38.1890	38.5827	38.9764
1.00	39.3701	39.7638	40.1575	40.5512	40.9449	41.3386	41.7323	42.1260	42.5197	42.9134
1.10	43.3071	43.7008	44.0945	44.4882	44.8819	45.2756	45.6693	46.0630	46.4567	46.8504
1.20	47.2441	47.6378	48.0315	48.4252	48.8189	49.2126	49.6063	50.0000	50.3937	50.7874
1.30	51.1811	51.5748	51.9685	52.3622	52.7559	53.1496	53.5433	53.9370	54.3307	54.7244
1.40	55.1181	55.5118	55.9055	56.2992	56.6929	57.0866	57.4803	57.8740	58.2677	58.6614
1.50	59.0551	59.4488	59.8425	60.2362	60.6299	61.0236	61.4173	61.8110	62.2047	62.5984
1.60	62.9921	63.3858	63.7795	64.1732	64.5669	64.9606	65.3543	65.7480	66.1417	66.5354
1.70	66.9291	67.3228	67.7165	68.1102	68.5039	68.8976	69.2913	69.6850	70.0787	70.4724
1.80	70.8661	71.2598	71.6535	72.0472	72.4409	72.8346	73.2283	73.6220	74.0157	74.4094
1.90	74.8031	75.1969	75.5906	75.9843	76.3780	76.7717	77.1654	77.5591	77.9528	78.3465
2.00	78.7402	79.1339	79.5276	79.9213	80.3150	80.7087	81.1024	81.4961	81.8898	82.2835
2.10	82.6772	83.0709	83.4646	83.8583	84.2520	84.6457	85.0394	85.4331	85.8268	86.2205
2.20	86.6142	87.0079	87.4016	87.7953	88.1890	88.5827	88.9764	89.3701	89.7638	90.1575
2.30	90.5512	90.9449	91.3386	91.7323	92.1260	92.5197	92.9134	93.3071	93.7008	94.0945
2.40	94.4882	94.8819	95.2756	95.6693	96.0630	96.4567	96.8504	97.2441	97.6378	98.0315
2.50	98.4252	98.8189	99.2126	99.6063	100.0000	100.3937	100.7874	101.1811	101.5748	101.9685
2.60	102.3622	102.7559	103.1496	103.5433	103.9370	104.3307	104.7244	105.1181	105.5118	105.9055
2.70	106.2992	106.6929	107.0866	107.4803	107.8740	108.2677	108.6614	109.0551	109.4488	109.8425
2.80	110.2362	110.6299	111.0236	111.4173	111.8110	112.2047	112.5984	112.9921	113.3858	113.7795
2.90	114.1732	114.5669	114.9606	115.3543	115.7480	116.1417	116.5354	116.9291	117.3228	117.7165
3.00	118.1102	118.5039	118.8976	119.2913	119.6850	120.0787	120.4724	120.8661	121.2598	121.6535
3.10	122.0472	122.4409	122.8346	123.2283	123.6220	124.0157	124.4094	124.8031	125.1969	125.5906
3.20	125.9843	126.3780	126.7717	127.1654	127.5591	127.9528	128.3465	128.7402	129.1339	129.5276
3.30	129.9213	130.3150	130.7087	131.1024	131.4961	131.8898	132.2835	132.6772	133.0709	133.4646
3.40	133.8583	134.2520	134.6457	135.0394	135.4331	135.8268	136.2205	136.6142	137.0079	137.4016
3.50	137.7953	138.1890	138.5827	138.9764	139.3701	139.7638	140.1575	140.5512	140.9449	141.3386
3.60	141.7323	142.1260	142.5197	142.9134	143.3071	143.7008	144.0945	144.4882	144.8819	145.2756
3.70	145.6693	146.0630	146.4567	146.8504	147.2441	147.6378	148.0315	148.4252	148.8189	149.2126
3.80	149.6063	150.0000	150.3937	150.7874	151.1811	151.5748	151.9685	152.3622	152.7559	153.1496
3.90	153.5433	153.9370	154.3307	154.7244	155.1181	155.5118	155.9055	156.2992	156.6929	157.0866
4.00	157.4803	157.8740	158.2677	158.6614	159.0551	159.4488	159.8425	160.2362	160.6299	161.0236
4.10	161.4173	161.8110	162.2047	162.5984	162.9921	163.3858	163.7795	164.1732	164.5669	164.9606
4.20	165.3543	165.7480	166.1417	166.5354	166.9291	167.3228	167.7165	168.1102	168.5039	168.8976
4.30	169.2913	169.6850	170.0787	170.4724	170.8661	171.2598	171.6535	172.0472	172.4409	172.8346
4.40	173.2283	173.6220	174.0157	174.4094	174.8031	175.1969	175.5906	175.9843	176.3780	176.7717
4.50	177.1654	177.5591	177.9528	178.3465	178.7402	179.1339	179.5276	179.9213	180.3150	180.7087
4.60	181.1024	181.4961	181.8898	182.2835	182.6772	183.0709	183.4646	183.8583	184.2520	184.6457
4.70	185.0394	185.4331	185.8268	186.2205	186.6142	187.0079	187.4016	187.7953	188.1890	188.5827
4.80	188.9764	189.3701	189.7638	190.1575	190.5512	190.9449	191.3386	191.7323	192.1260	192.5197
4.90	192.9134	193.3071	193.7008	194.0945	194.4882	194.8819	195.2756	195.6693	196.0630	196.4567
5.00	196.8504	197.2441	197.6378	198.0315	198.4252	198.8189	199.2126	199.6063	200.0000	200.3937

Metric surface finishes are converted the same way as any other metric–to-English conversion. A common finish found on metric drawings is 3.2. The table gives an equivalent of 125.9843. A 125 finish is common, usually indicating a machined surface.

Measuring Units

Unit A: Pages 2652–2655 Symbols and Abbreviations

Letters of the Greek alphabet are frequently used in mathematical formulas. Greek letters and standard abbreviations are found on pages 2652–2655.

Index navigation paths and key words:

— ANSI/abbreviations/pages 2652–2655
— abbreviations/mathematical signs/page 2654
— abbreviations/welding/page 1479
— mathematical/signs and abbreviations/page 2654
— drawing/symbols/geometric/page 612

Navigation Hint: For more information on symbols, see Section 4; Unit E, Dimensioning, Gaging, and Measuring; in *Machinery's Handbook Made Easy* pages 51–52.

ASSIGNMENT

List the key terms and give a definition of each.

Meter	Minute
Millimeter	Second
Degree	Roughness Average

APPLY IT!

1. What device is used to inspect surface finishes?

2. What is the decimal equivalent of 15/16?

3. Convert 4 1/2 to millimeters.

4. Convert 45 seconds to decimal degrees

5. What is 12° 16′ 56″ in decimal degrees?

6. How many millimeters are in a meter?

12

Answer Keys

SECTION 1

Answer Key for Assignment

Numbers	Whole numbers are 0, 1, 2, 3, 4, 5, 6, 7, 8, 9
Fraction	A value separated by a line consisting of a numerator and a denominator
Improper Fraction	A fraction where the numerator is greater than the denominator
Decimal	A value expressed in terms of tenths, hundredths, and thousandths
Reciprocal	The number obtained by dividing 1 by the given number
Numerator	The top number of a fraction
Mixed Number	A value consisting of a whole number and a fraction
Denominator	The bottom number of a fraction
Proposition	As in geometry, a statement that is known to be true
Equilateral	A triangle where all three sides are equal
Isosceles	A triangle where two sides are equal
Hypotenuse	The longest side of a right triangle
Bisect	To divide an angle into two equal parts
Perpendicular	At ninety degrees to a feature such as a line or plane
Sine	The length of the opposite side divided by the length of the hypotenuse
Cosecant	The reciprocal of sine
Cosine	The length of the adjacent side divided by the length of the hypotenuse
Secant	The reciprocal of cosine
Tangent	The length of the opposite side divided by the length of the adjacent side
Cotangent	The reciprocal of tangent
Reciprocal	Multiplicative inverse, i.e. the reciprocal of x = 1/x

Answer Key for *Apply it!* Part 1

1. a. 30°
   b. 30°
   c. 30°
   d. 150°
   e. 150°

Answer Keys

2. X = 55°
3. X = 51°
4. X = 114°; Y = 91°
5. X = 10°; Y = 60°
6. X = 115°
7. X = 140°
8. X = 31°

Answer Key for *Apply it!* Part 2

1. .55919
2.

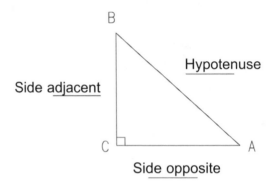

3. 1.73205
4. 1/2
5. .74314
6. 1.05146222
7. a. 18.43494
   b. 71.56506
   c. 3.16227
8. a. 41.81
   b. 48.19

SECTION 3

Answer Key to Assignment

Alloy:	Two or more elements forming a mixture in a solid solution
Element:	A pure chemical substance consisting of one type of atom
Hardened:	A condition in which a metal is resistant to penetration
Tempered:	A condition in which internal stresses have been relieved
Pig Iron:	An intermediate ingredient in the steel-making process
Anneal:	A hot process that softens a metal

Stress Relieve:	To reduce the internal stresses in a metal commonly caused by machining
Thermosets:	The type of plastic that forms chemical bonds when heated; this type of resincannot be reprocessed
Thermoplastics:	The type of plastic that can be repeatedly melted by heating and solidified by cooling

SECTION 4

Answer Key to Assignment

Meter	The standard of measurement used in the metric system
Millimeter	One thousandth of a meter
Inch	A standard unit of measurement used in the "English" system
Force Fit	The assembly of mating parts when the female component is larger than the male component
Visible Line	The lines on an engineering drawings that represent the outside shape of the part; also called Object Lines
Hidden Line	Short, dashed lines on an engineering drawing that represent interior detail of the part
Center Line	Lines on an engineering drawing that represent the axis of a circular feature
Datum	A feature that is used as a starting point or a place from which dimensions begin
Basic Dimension	A numerical value used to describe the theoretically exact size, orientation, location, or profile of a feature or datum
Interference Fit	Another term for Force Fit also known as "Press Fit"
Tolerance	Allowable deviation from a specified dimension
Discrimination	The degree to which an instrument divides units of measurement
Maximum Material Condition	The condition in which a feature contains the most material. A hole is at its Maximum Material Condition (MMC) when it is the smallest because that is when it contains the most material. A shaft is at its MMC when it is at its largest size.

Answer Key to *Apply It!*

1. One revolution = .025 = 1/40, so 40 revolutions = 1 inch; answer: 40
2. a. .059
   b. 1.043
   c. 1.483

Answer Keys

3. a. .741
    b. .774
    c. .839
    d. .7187
4. 1,000 or 10,000
5. a. 125   drilling
    b. 16   barrel finishing
    c. 250   sawing
    d. 8   grinding
    e. 63   milling
6. a. 1/2″ O.D. bushing   .5017 − .5007 = .001
    b. 5/16″ O.D. bushing   .3141 − .3132 = .0009
    c. 7/8″ O.D. bushing   .8768 − .8757 = .0011
7. a. 3/8″ O.D. bushing   .3763 − .3760 = .0003
    b. 7/16″ O.D. bushing   .4389 − .4385 = .0004
    c. 1 1/2″ O.D. bushing 1.0015 − 1.0010 = .0005
8. Instructor: Grade the sketch using proper techniques from page 609 MHB 28 table 2.
9. n/a

SECTION 5

Answer Key to Assignment

Positive rake	The inclination of the tool face pointing up at the tip of the tool
Negative rake	The inclination of the tool face pointing down at the tip of the tool
Nose radius	The rounded point of the cutting edge of the tool
High-speed steel	A super-alloy cutting tool material
Reaming allowance	The reamer diameter minus the size of the previously drilled hole
Helical flute	A reamer that has spiral cutting edges
Shank	The part of the reamer that is used to drive the tool
Point angle	The angle included in the cutting lips of the drill
Margin	Adjacent to the top of the cutting lips
Tapered shank	Conical shank used to drive larger diameter drills
Flute	Spiral grooves cut in the body of the drill
Effective thread	The depth of useable, full threads
Pitch	The distance from one part of a thread to the corresponding point of the next thread
Major diameter	The outside diameter of a male thread; also called crest

Answer Keys

Minor diameter   The root of the thread

Blind hole   A hole that does not go through

Cratering   The condition of a cutting tool where a cavity-like depression develops just behind the cutting edge

Answer Key for *Apply it!* Part1

1. Square: 8; Triangular: 6
2. T triangular
   N 0 degrees
   L .002 to .005 tolerance
   A with mounting hole
   4 4/8 (1/2) inscribed circle
   3 3/16 thickness
   2 2/64 (1/32) nose radius
   No position 8
   No position 9
   B honed .003 to .005
3. a. round
   b. round or square
   c. triangular
   d. triangular
4. Three to five times greater. (750 to 1250 RPM).
5. With a stud through a mounting hole in the insert (pin lock); with a top clamp; with a machine screw; brazed (welded)
6. Triangular

Answer Key for *Apply It!* Part 2

Reamer Size	.2500	.3125	.3750
Drill #1	15/64 (.234)	9/32 (.281)	T (.358)
Drill #2	C (.2420)	N (.302)	U (.368)

There is more than one correct answer for this table. It is important that Drill #2 complies with the Reaming Allowance chart in this unit. The choice for Drill #1 should be about .010 to .015 less than Drill #1 and is subject to available drill sizes.

Answer Key for *Apply It!* Part 3

1. .375
2. #26
3. Shank
4. From .234 to .413
5. 1/64
6. 10.262 mm

Answer Keys

Answer Key for *Apply It!* Part 4

1. 7.0 mm = .2755; J drill (.2770)
2. .2340 to .4130
3. 3/8 (.375); 1/16 (.0625)
4. Metric system begins with "M"; pitch is given rather than number of threads in a distance.
5. a. .750 to 1.00
   b. 12 mm to 16 mm (.472 to .629)
   c. .5625 to .750
6. a. #7 (.2010)
   b. Letter F (.257)
   c. Letter Q (.332)
   d. 5/16 (.3125)
   e. 29/64 (.4531)

SECTION 6

Answer Key for Assignment

High Speed Steel	High alloy tungsten or molybdenum tool steel with the ability to withstand high temperatures and abrasion and retain their hardness deeply into the cutting tool.
RPM	Revolutions per minute
Tool Steel	Alloy steel used for making tools
Cutting Speed	The number of feet that pass the cutting tool in one minute
Thermal Shock	The result of rapid heating and cooling
Plain carbon steel	Iron and less than 2% carbon
Ferrous and Non-Ferrous	Ferrous metals contain iron; non-ferrous metals do not contain iron.
Alloy Steel	Steel with alloying elements other than carbon
Cratering	The tendency of a material to weld itself to the cutting tool, break off, and cause the tool to chip in a crater-like shape
Surface Feet per Minute	Also known as cutting speed, the number of feet that pass the cutting tool in one minute; a measurement of machinability

Answer Key for *Apply It!*

1. Cutting speed = 300; feed = .012 ipr found on the bottom of page 1000 in a general note; RPM: $4 \times 300 \div 4.50 = 266.6$
2. Cutting speed = 75; feed = .012; RPM: $4 \times 75 \div .375 = 800$

Answer Keys

3. Cutting speed = 60; RPM = 815; for 15/16: RPM = 234
4. Milling: ipm (inches per minute); Turning: ipr (inches per revolution)
5. Reference page 1198: Work not flat, burnishing of work, burning or checking, chatter marks, wheel loading, wheel glazing
6. 36 to about 220
7. C: Abrasive Silicon carbide; 46: 46 Grit; J: Grade J medium; 6: Structure 6, medium; V: bond, Vitrified bond
8. Reference page 1198: Grade too soft, dirty coolant, diamond loose or chipped, no or poor magnetic force, Chuck surface worn or burred, insufficient blocking of parts, loose dirt under guard.
9. M codes:

Table 3. M-codes

M code	Description	M code	Description
M00	Program stop	M09	Coolant OFF
M01	Optional program stop	M19	Spindle orientation
M03	Spindle rotation normal (clockwise)	M30	Program end
M04	Spindle rotation reverse (counterclockwise)	M60	Automatic Pallet Change (APC)
M05	Spindle stop	M98	Subprogram call
M06	Automatic Tool Change (ATC)	M99	Subprogram end
M08	Coolant ON		

10. G01 Linear Interpolation
11. Feed rate = RPM × Feed per Tooth × # of cutting flutes, so 7.68 = 480 × (Feed Per Tooth) × 4
    Feed per Tooth is equal to the chip thickness = .004
12. .002
13. .015 × 2 = .030
14. Multiply the RPM by the ipr

Refer to Table 1 and Table 2 for the following questions
15. ipr for a for a "C" drill (.242) is .003; RPM is not needed
16. ipr for a "U" (.368) is .004
17. Feeds expressed in "inches per minute" are commonly found on milling machines

Table 1

Feed in inches per revolution for drills	
0-1/8	.001
1/8-1/4	.003
1/4-3/8	.004
3/8-1/2	.006
1/2-1.0	.010

249

Answer Keys

Table 2

Workpiece Material	Cutting Tool Material	
	HSS	Carbide
Aluminum	300	1500
Mild Steel (1018)	90	300
Alloy Steel (4140; 4340)	60	240
High Carbon Steel (1040; 1060)	50	150
Stainless Steel	50	150

18. RPM $\times$ feed per tooth $\times$ # of flute on the cutter = ipm
19. RPM = 4 $\times$ 1500 $\div$ 3.0 = 2000
20. RPM = 4 $\times$ 60 $\div$ .25 = 960
21. RPM = 4 $\times$ SFM $\div$ dia. so 100 = 4 $\times$ (SFM) $\div$ 6; SFM = 150
22. Feed = RPM $\times$ feed per tooth $\times$ # of flutes on the cutter
23. 100 $\times$ .006 $\times$ 8 = 4.8
24. 100 $\times$ .006 $\times$ 6 = 3.6
25. RPM = 4 $\times$ 50 $\div$ .22 = 909; ipm = 909 $\times$ .004 $\times$ 2 = 7.272
26. RPM = 4 $\times$ 50 $\div$ .200 = 1000; so ipm = 1000 $\times$ .004 $\times$ 4 = 16
27. 4 $\times$ 90 $\div$ 6.0 = 60
28. ipm = 60 $\times$ .006 $\times$ 24 = 8.64
29. .006
30. 4 $\times$ 1500 $\div$ 1.0 = 6000

SECTION 8

Answer Key for Assignment

Nominal: Standard size from which dimensions are derived

Blind Hole: A hole that does not go through the material

Press Fit: When one component is larger than its mating part; also known as interference fit

Slip Fit: When an allowance is designed into mating parts

Hexagon Head: With respect to socket head fasteners; a six-sided impression for a hex key

Bolt: A threaded fastener that is held into an assembly with a nut

Screw: A fastener that threads into and fits into a tapped hole to hold an assembly

Pilot Diameter: The non-cutting guiding portion of a cutting tool

Answer Keys

Counterbore:	A hole with a stepped diameter to accommodate a fastener head; also the tool that creates the recessed portion of a fastener hole for the head of a fastener
Countersink:	A tapered hole made to accept a flat head fastener; or the tool that creates the tapered or chamfered portion of hole

Answer Key for *Apply It!*

1. A *bolt* uses a nut and a *screw* is threaded into a tapped hole
2. Flat head fasteners tightly locate components together by the locking angle of the head of the screw. Use where no movement or adjustment is allowed.
3. The length of engagement for a dowel is two times it diameter.
   a. $1/2'' \times 2 = 1.0$
   b. $.3750 \times 2 = 3/4$
   c. $1/4'' \times 2 = 1/2$
   d. $.1875 \times 2 = 3/8$
4. The minimum thread engagement for the following thread sizes is 1½ times its diameter.
   a. #10 = .19 $.19 \times 1.5 = .285$
   b. $.50 \times 1.5 = .75$
   c. $3/8 \times 1.5 = .562$
   d. $3/8 \times 1.5 = .562$
5. Most dowels are four times their length because they should contact the component for a length of two times the dowels diameter in both parts.
6. The counterbore depth for the following socket head cap screws is their thread diameter (which is the same as their head height) plus .03 for clearance
   a. $3/8 + .03 = .405$
   b. $1/2 + .03 = .53$
   c. $1/4 + .03 = .28$
   d. #10 =.19 $.19 \times 1.5 = .22$

SECTION 9

Answer Key for Assignment

Chamfer	Conical lead at the start of the thread
Crest	The surface of the thread that is the farthest from the centerline of the thread
Thread Series	Accepted groups of diameter/pitch combinations
Unified National Coarse	A thread series also known as UNC
Unified National Fine	A thread series also known as UNF

Answer Keys

Minor Diameter	The surface of the thread that is the closest to the centerline of the thread
Major Diameter	The outside diameter of the thread also known as the crest
Nominal Size	The general size used to identify a feature
Pitch Diameter	An imaginary cylinder halfway between the crest and root of the thread
Root	The surface of the thread that is closest to the centerline of the thread

Answer Key for *Apply It!* Part 1 (Drawing #123–456)

Use Table 3 on page 1817 for questions 1 and 2.

1. 0.9980/0.9755
2. 0.8734/0.8579
3. $1 - 8 = 0.125$; $7/8 - 14 = 0.0714$
4. Unified National Coarse
5. $1 - 8 = 8 \times 1.5 = 12$; $7/8 - 14 = 14 \times 1.5 = 21$
6. 1018 CRS (cold rolled steel)
7. 1/8 or .125
8. 0.8270/0.8189
9. It is the funtional part of the thread
10. 2 – general purpose fit; A – external

Answer Key for *Apply It!* Part 2 (Drawing #123–457)

1. M: metric; 8: 8mm external nominal diameter; 1.25: pitch; 6g: tolerance grade and position
2. 60 degrees
3. 8mm nominal diameter; 10mm nominal diameter.
4. 16mm =0. 6299 so 7/8 or 1.0 inch
5. $20 \times 1.5 = 30$

Answer Key for *Apply It!* Part 3 (Tap drill questions)

1. .293 (M drill)
2. .117 (32 drill)
3. $1/2 - 13$: 1.0 deep $5/16 - 18$: .625

$$\text{outside dia.} = \frac{-0.01299 \times \% \text{ of thread}}{\# \text{ of threads per inch}} = \text{tap drill}$$

SECTION 11

Answer Key for Assignment

Keyseat	The groove in a shaft that accepts a key.
Keyway	In an assembly with a shaft and a hub, the groove in the hub.
Woodruff key	A semi-circular key that fits into a semi-circular groove that is shallower than the key.
Broach	A toothed cutting tool that is pushed or pulled through a finished bore to produce a keyway.
Keyseating machine	A machine that cuts *keyways* in hubs, gears, and pulleys by drawing or pushing a *broach* through a finished bore (hole).
Wire EDM	An electrical manufacturing process that uses a wire (.002 to .010) that follows a programmed path to machine an accurate shape.
Keyseat alignment tolerance	The amount of acceptable error between the centerline of the key seat and the centerline of the shaft or bore.
Clearance fit	A designed allowance between mating components.
Interference fit	Condition when designed mating parts are such that the male component is larger than the female component, causing a force fit.

Answer Key for *Apply It!*

1. a. 1/8 × 1/8
   b. 3/16 × 3/16 or 1/4 × 1/4
   c. 5/16 × 5/16
   d. 5/8 × 5/8
2. a. 1/16 × 1/4
   b. 3/16 × 5/8
   c. 3/8 × 1.0
   d. 3/8 × 1/4
3. a. .125
   b. .1562
   c. .0937
4. a. 1/2″
   b. 1/2″
   c. 1/2″
5. Rectangular keys are recommended on shafts 6 1/2″ diameter and larger.

Answer Keys

SECTION 12

Answer Key for Assignment

Meter	The metric unit of measurement
Millimeter	One thousandth of a meter
Degree	A unit of angular measurement
Minute	One 60th of a degree
Second	One 60th of a minute
Roughness Average	An average of the peaks and valleys observed in the inspection of a surface

Answer Key for *Apply It!*

1. The profilometer
2. .9375
3. 114.3mm
4. .75 degrees
5. 12.2822222
6. 1000

In Conclusion...

The breadth and depth of wisdom contained within *Machinery's Handbook* is unmatched. Throughout this book, I have strived to provide you with a comprehensive, easy-to-use guide that I hope will help you to unlock the wisdom in *Machinery's Handbook,* making your job easier and more enjoyable. However, there are some things that cannot be captured in the words of a book, no matter how detailed it may be. It is these intangible qualities that make the professions of those of us who study these pages as much art as they are science.

It is for this reason that I hope you will consider stepping away from your reference books from time to time and tapping into the rich knowledge base which undoubtedly surrounds you. I'm talking, of course, about your colleagues, teachers, students, peers, apprentices, and superiors. Too often, we allow priceless knowledge to disappear with the retirement of a seasoned colleague. At the end of the day, we all have an obligation not only to share in the knowledge that comes from our own experience, but to reach out and let the people around us know that their unique perspectives are deeply valued.

Congratulations on your pursuit of excellence in the art of metal removal. Equipped with *Machinery's Handbook*, this book, a strong network of coworkers, and an open mind, I am confident you will find success in anything you put your mind to.

Edward Janecek

Glossary

Alloy	Two or more elements forming a mixture in a solid solution
Alloy Steel	Steel with alloying elements other than carbon
Aluminum Oxide	Also known as corundum, an abrasive used extensively in industry and used in the production of aluminum
Annealing	A hot process that softens a metal
Austenitic	A solid solution of iron and carbon or iron carbide
Basic Dimension	A numerical value used to describe the theoretically exact size, orientation, location, or profile of a feature or datum
Bisect	As in geometry, to divide an angle into two equal parts
Blind Hole	A hole that does not go through the material
Bolt	A threaded fastener that is held into an assembly with a nut
Broach	A toothed cutting tool that is pushed or pulled through a finished bore to produce a keyway or special shape
Carbon Steel	Steel consisting of iron and carbon with no other alloying elements
Center Line	Lines on an engineering drawing that represent the axis of a circular feature
Chamfer	Conical lead at the start of a thread or end of a shaft or edge
Chromium	A common alloying element in steel; ingredient that makes stainless steel corrosion resistant
Clearance Fit	A designed allowance between mating components
CNC Machining	Automated machine tools used in manufacturing that follow a pre-programmed path
Cosecant	The reciprocal of sine
Cosine	In a right triangle, the length of the adjacent side divided by the length of the hypotenuse
Cotangent	The reciprocal of tangent
Counterbore	A hole with a stepped diameter to accommodate a fastener head; also the tool that creates the recessed portion of a fastener hole for the head of a screw
Countersink	A tapered hole made to accept a flat head fastener; or the tool that creates the lead or chamfered portion of a hole
Cratering	The tendency of a material to weld itself to the cutting tool, break off, and cause the tool to chip in a crater-like shape
Crest	The surface of the thread that is the farthest from the centerline of the thread
Cubic Boron Nitride (CBN)	Next to diamond, the hardest known material. Used in cutting tool material for hard and tough materials
Cutting Speed	Cutting speed is the number of feet that pass the cutting tool in one minute

Datum	A feature that is used as a starting point or a place from which dimensions begin
Decimal	A value expressed in terms of tenths, hundredths, and thousandths
Degree	A unit of angular measurement
Denominator	The bottom number of a fraction
Diamond	The hardest known material. Used in cutting tool material for tough, hard metals
Discrimination	The degree to which an instrument divides units of measurement
Dowel	A precision cylindrical component commonly used to locate features in an assembly
Effective Thread	The depth of useable, full threads
Electrical Discharge Machining	A machining operation that uses rapidly recurring sparks to vaporize the electrically conductive workpiece with an electrode or wire
Element	A pure chemical substance consisting of one type of atom
Equilateral	If three sides of a triangle are equal they are said to be equilateral
Ferrite	Scientific term for iron. Solid solution with iron as the main ingredient
Ferrous & Non-Ferrous	Ferrous metals contain iron; non-ferrous metals do not contain iron
Fluid Mechanics	The area of manufacturing that studies fluid and forces
Flute	Spiral grooves cut in the body of a cutting tool
Force Fit	The assembly of mating parts when the female component is larger than the male component
Fractions	A value separated by a line consisting of a numerator and a denominator
G-Code	The programming language used to control a computer numerically controlled (CNC) machine
Hardened	A condition in which a metal is resistant to penetration. A material that has gone through the process of hardening by heat treating is said to be hardened
Helical Flute	A cutter with spiral cutting edges
Hexagon Head	With respect to socket head fasteners; a six-sided impression for a hex key
Hidden Line	Short, dashed lines on an engineering drawing that represent interior detail of the part
High Speed Steel	High alloy tungsten or molybdenum tool steel with the ability to withstand high temperatures and abrasion
Hot Hardness	The ability of a material to retain its cutting edge or resist deforming at elevated temperatures
Hypotenuse	The longest side of a right triangle
Improper Fraction	A fraction where the numerator is greater than the denominator
Inch	A standard unit of measurement used in the "English" system
Interference Fit	Another term for Force Fit also known as "Press Fit" in which one component is larger than the mating component, resulting in an "interference"

258

Isosceles	A triangle where two sides are equal
Keyseating Machine	A machine that cuts *keyways* in hubs, gears, & pulleys by drawing or pushing a *broach* (special cutter) through a finished bore (hole)
Keyseat	The groove in a shaft that accepts a key
Keyway	In an assembly with a shaft and a hub, the *keyway* is the groove in the hub
Laser	An acronym for "Light Amplification by the Stimulated Emission of Radiation". In manufacturing lasers are used to cut, engrave, heat treat, and weld.
Major Diameter	The outside diameter of a thread also known as the crest
Margin	Adjacent to the top of the cutting lips of a drill
Martensite	A crystalline structure of hardened steel
Maximum Material Condition (MMC)	The condition in which a feature contains the most material. A hole is at its Maximum Material Condition (MMC) when it is the smallest because that is when it contains the most material. A shaft is at its MMC when it is at its largest size.
M-Code	Machine codes used in CNC machining centers to control machine functions such as spindle start
Metal	A material that reflects light and conducts electricity
Meter	The standard of measurement used in the metric system
Millimeter	One thousandth of a meter
Minor Diameter	The surface of a thread that is the closest to the centerline of the thread
Major Diameter	The outside diameter of the thread also known as the crest
Minute	One 60th of a degree
Mixed Number	A value consisting of a whole number and a fraction
Negative Rake	The inclination of the tool face pointing down at the tip of the tool
Nominal	Standard size that dimensions are derived from
Nominal Size	The general size used to identify a feature
Non-ferrous	A metal that does not contain iron
Nose Radius	The rounded point of the cutting edge of the tool
Numbers	Whole numbers are 0, 1, 2, 3, 4, 5, 6, 7, 8, 9
Numerator	The top number of a fraction
Perpendicular	At ninety degrees to a feature such as a line or plane
Pig Iron	Semi-refined iron used in the steel making process
Pilot Diameter	The non-cutting guiding portion of a cutting tool
Pitch	The distance from one part of a thread to the corresponding point of the next thread
Pitch Diameter	An imaginary cylinder halfway between the crest and root of the thread
Plain Carbon Steel	Iron and less than 2% carbon
Point Angle	The angle included in the cutting lips of a drill
Positive Rake	The inclination of the tool face pointing up at the tip of the tool

Press fit	When one component is assembled to a smaller mating component; also known as interference fit
Proposition	As in geometry, a statement that is known to be true
Reamer	A cutting tool used in industry to slightly enlarge a previously drilled hole
Reaming Allowance	The reamer diameter minus the size of the previously drilled hole
Reciprocal	The number obtained by dividing 1 by the given number
Resinoid	A type of bonding agent used in grinding wheels
Root	The surface of the thread that is closest to the centerline of the thread
Roughness Average	An average of the peaks and valleys observed in the inspection of a surface
RPM	Revolutions per minute
Screw	A fastener that is intended to be used in a threaded hole
Secant	The reciprocal of cosine
Second	One 60th of a minute
Set Screw	A headless fastener commonly used to secure shafts and lock machine components
Shank	The part of the cutter that is used to drive the tool
Shear Strength	The ability of a material to resist cutting forces
Shoulder Bolt	See stripper bolt
Silicon Carbide	As in grinding wheels, the type of manufactured abrasive that is commonly used to grind non-ferrous metals, cast iron and tungsten carbide
Sine	In a right triangle, the length of the opposite side divided by the length of the hypotenuse
Slip fit	A condition where there is an allowance designed into mating parts
Soldering and Brazing	A relatively low temperature process used to join metals
Statistical Process Control (SPC)	Using statistical methods to monitor, control, and predict the quality and output of a manufacturing process
Stainless Steel	A corrosion-resistant steel alloy that contains a minimum of 11% chromium
Stress Relieve	To reduce the internal stresses in a metal commonly caused by machining or welding
Stripper Bolt	A fastener with a ground diameter adjacent to the threads; also known as a shoulder bolt
Surface Feet per Minute	Also known as cutting speed, the number of feet that pass the cutting tool in one minute. A measurement of machineability
Tangent	In a right triangle, the length of the opposite side divided by the length of the adjacent side
Tapered Shank	Conical shank used to drive larger diameter drills
Temper	A hot process used in industry to increase toughness and decrease brittleness
Tempered	A condition in which internal stresses of a metal have been relieved

Tensile Strength	Stress measured as force per unit area as in pounds per square inch
Thermal Shock	The result of rapid heating and cooling
Thermoplastics	The type of plastic that can be repeatedly melted by heating and solidified by cooling
Thermosets	The type of plastic that forms chemical bonds when heated (cannot be reprocessed)
Thread Grinding	A precision method of producing threads on a grinding machine using a special grinding wheel that has the profile of the thread on the wheel
Thread Milling	Producing threads with a milling cutter in a helical path, usually on a CNC machine.
Thread Rolling	The process of producing threads by forming with a die pressed against a blank
Thread Series	Accepted groups of diameter/pitch combinations
Tolerance	Allowable deviation from a specified dimension
Tool Steel	Alloy steel used for making tools
Torsional Strength	The ability to resist twisting forces
Toughness	The ability to resist impact
Unified National Coarse	A thread series also known as UNC
Unified National Fine	A thread series also known as UNF
Visible line	The lines on an engineering drawings that represent the outside shape of the part; also called Object Lines
Wire EDM	An electrical manufacturing process that uses a .002 to .010 diameter wire which follows a programmed path to machine an accurate shape
Woodruff Key	A semi-circular key that fits into a semi-circular groove that is shallower than the key

Index

Index

Index

Index